The Latin American Studies Book Series

Series Editors

Eustógio W. Correia Dantas, Departamento de Geografia, Centro de Ciências, Universidade Federal do Ceará, Fortaleza, Ceará, Brazil

Jorge Rabassa, Laboratorio de Geomorfología y Cuaternario, CADIC-CONICET, Ushuaia, Tierra del Fuego, Argentina

Andrew Sluyter, Louisiana State University, Baton Rouge, LA, USA

The Latin American Studies Book Series promotes quality scientific research focusing on Latin American countries. The series accepts disciplinary and interdisciplinary titles related to geographical, environmental, cultural, economic, political and urban research dedicated to Latin America. The series publishes comprehensive monographs, edited volumes and textbooks refereed by a region or country expert specialized in Latin American studies.

The series aims to raise the profile of Latin American studies, showcasing important works developed focusing on the region. It is aimed at researchers, students, and everyone interested in Latin American topics.

Submit a proposal: Proposals for the series will be considered by the Series Advisory Board. A book proposal form can be obtained from the Publisher, Juliana Pitanguy (juliana.pitanguy@springer.com).

More information about this series at http://www.springer.com/series/15104

Fabiana Lopes da Cunha · Jorge Rabassa
Editors

Festivals and Heritage in Latin America

Interdisciplinary Dialogues on Culture, Identity and Tourism

 Springer

Editors
Fabiana Lopes da Cunha
Paulista State University "Julio de Mesquita
Filho"—UNESP
São Paulo, Brazil

Jorge Rabassa
National Academy of Sciences Argentina
CADIC-CONICET
Ushuaia, Tierra del Fuego, Argentina

ISSN 2366-3421   ISSN 2366-343X (electronic)
The Latin American Studies Book Series
ISBN 978-3-030-67984-2   ISBN 978-3-030-67985-9 (eBook)
https://doi.org/10.1007/978-3-030-67985-9

This Springer imprint is published by the registered company Springer Nature Switzerland AG
The registered company address is: Gewerbestrasse 11, 6330 Cham, Switzerland

To Rodrigo Donato (In Memorian)

To our friend, partner, great musician, who contributed a lot with UNESP/Ourinhos and with our events in cultural activities. I am sure that wherever you are, you are celebrating life with a lot of partying, good music and surrounded by enlightened people.

Preface

From the "Rites of Spring" to the "Roller Coaster Loop": Reflections on Heritage

As we approach the outskirts of Verdun on Route Nationale 3 from Metz, having already contemplated with pleasure the serenity of the rolling hills and meadows of the Vosges countryside and the disciplined guard of honour of robust oaks, we are suddenly surprised, a few kilometres from the city, by a gloomy view. A blur in the landscape. A cemetery. Piled up on the side of the road are crushed corpses, crumpled bodies, shimmering skeletons. But it is a cemetery without crosses, without tombstones, without flowers. There are few visitors. In general, travellers do not even notice the place. But it is an illustrious memorial of the twentieth century and of our cultural references. Many would say that it is a symbol of modern values and objectives, of our struggle and our remorse, the contemporary interpretation of Goethe's conjure, stirb und werde, "die and transmute yourself". It is a cemetery of cars.

If you continue to Verdun, cross the city and then take the north-east route by secondary roads, you can find the path which leads to a larger cemetery. This has crosses. Thousands of them. Rows and Rows. Symmetrical. White. All the same. More people pass through the car cemetery today than this one. More people identify themselves with the crushed cars than with the now impersonal horror which this cemetery evokes. This is the cemetery in memory of those who died during the battle of Verdun in the First World War (Eksteins: 1991, p. 11).

The "Rites of Spring", a book written still in the twentieth century, speaks of death and destruction, but also of "transmutation". It speaks of the connection between our concern with speed, movement, and, as in the struggle for freedom, we end up acquiring the power of final destruction, the Great War, the dance of death, hence the title, which links it to the presentation of this great artistic and musical work by Stravinsky, whose premiere was in Paris in May 1913, before the outbreak of the war that would change the form and meaning of our lives and our history.

And, the readers must be asking themselves: what would be the connection between this discussion and the themes treated in this book?

The first wave of folklore, concerned with collecting, preserving and building an "ideal" of a popular culture, as a heritage to be preserved, occurs in the Third French Republic. "It explores a rural world that the railway, the military service (let alone the mass media) had not yet put in contact with the city: a world that will move quickly after the First World War" (Certeau 2012, p. 63). This popular one that we find in the folklorist magazines of the time is associated with what would be natural, "the true, the naive, the spontaneous, the childhood" (Certeau 2012, p. 63). In other words, a construction that generalized and gave unity to a repertoire that disturbed part of the French and European elite of the period. At that time, popular culture was defined as both a historical and a geographical heritage. Tidy, "dominated", folklore would guarantee a vision and organization of the popular repertoire in a kind of "reassuring museum" (Certeau 2012, p. 63).

Sevcenko, in his "A Corrida Para o Século XXI: No loop da montanha-russa", brought up some more elements to think about these different moments in the history of the Western world and to reflect on what Gonçalves called "unease in the heritage". Sevcenko compared the emotions of being on a roller coaster, with all the exaggerations that it entails, with some of the most striking trends of our time. The first phase, from the sixteenth century to the middle of the nineteenth, would be of "continuous, methodical and persistent ascent" and would be related to the optimistic vision of the future linked to the formula "order and progress", which many believed to be of "abundance, rationality and harmony" (Sevcenko 2001, p. 14–5). The second, which coincides with the beginning of the first folkloristic wave, would be that of the vertiginous precipitation of the roller coaster after the slow and continuous climb. With the Scientific-Technological Revolution around 1870, it seemed that all the achievements linked to the unfolding of that revolution and the "confidence in progress" seemed to have reached its apex. And then, in an unexpected sudden, came the plunge into a vacuum, the chaotic and destructive spasm, the horror engulfed history: the outbreak of the Great War [...]" (Sevcenko 2001, p. 16). This period also coincides with the creation of the "Rites of Spring", which foresaw what was to come: mass destruction, now viable by broad technological resources. Such a destructive scale would only be overcome by World War II. Despite the resumption of scientific and technological development, "the sense of an impending apocalypse" prevailed. Not by chance, Heritage Letters began to be built and written from the 1930s onwards. The first focus is mainly on the cultural heritage of the most important nature.

> The third phase in our image of the roller coaster is the loop, the final and definitive syncope, the climax of precipitated acceleration, under whose extreme intensity we relax our impulse to react, delivering the numb points, resignedly accepting to be driven to the end by the titanic machinery. This stage would represent the current period, marked by a new dramatic surge of transformations, the Microelectronics Revolution. The scale of the changes unleashed from that moment on is of such magnitude that the two previous moments seem like projections in slow motion (Sevcenko 2001, p. 16).

We believe that this scenario has stimulated debate, actions and extensive literary production on the theme of "cultural heritage".

> A historian in the year 2115 will probably wonder why people, at the turn of the twentieth century to the twenty-first, were so sensitive to the words "heritage" and "memory" and so obsessed with actions to protect and preserve their "cultural assets". What threat haunted their hearts and minds? One might also wonder how these conservationist attitudes could live with such devastating natural and social catastrophes as those we know in the contemporary world and which affect precisely those objects that are the target of tireless protection and preservation efforts. How can we understand the logic of these destructive actions over which, apparently, effective collective control could not be exercised?

Nowadays, any material object, any space, any social practice, any type of knowledge can be identified, celebrated or contested as "patrimony" by one or more social groups (Gonçalves 2015, p. 212).

Just as the "abuse of memory" has already been discussed, we must also talk about how there is a certain inflationary character in what we today call "patrimony". But it is important to think of such debates and actions in this sense as what "symptoms" are

of our experiences of time: by describing and analysing their historical and geographical variations, we would actually be comparing different ways of experiencing time (Hartog 2003). For him, the extraordinary expansion of heritage in the contemporary world must be understood as the symptom of a crisis in the ways we experience the relationships between past, present and future (Gonçalves 2015, pp. 216–7).

We also know that heritage actions serve the objectives of the tourism industry on a global scale,

strategies for the construction of "identities", the formation of individual and collective subjectivities, political and economic claims by social groups, or state policies (Gonçalves 2015, p. 218).

This expansion of heritage in the contemporary world and the contradiction between a certain preservationist "obsession" and the rage of destruction that is being wrought on the most diverse cultural goods reveal the two sides of the same issue. This outburst of the various forms of cultural heritage which part of society is committed to preserving and rebuilding reveals not only the effort to recognise social "identities", but also the "supposed risks of their 'loss'" (Gonçalves 2015, p. 218). Like the last movement of the roller coaster loop, where we come out of the experience stunned and "apatet", technological acceleration and the voracious acceleration of change make "the universe of possibilities and expectations [...] more and more unpredictable, irresistible and incomprehensible" (Sevcenko 2001, pp. 16–17). And, if part of society tends to leave to think of the damage then, on the other hand, "it invades the present in the form of 'patrimonies'". The collection of objects, records, which are collected, organized and exhibited in museums and other spaces to be consumed would be what Hartog calls an "eternal gift". As Gonçalves stated

It is likely that a collective work of mediating and balancing contradictions is at stake in our contemporary way of representing time, a conception in which the future no longer shines as the focus of utopian hopes, and the past is preserved or rebuilt in the vain expectation of stopping time (Gonçalves 2015, p. 218).

In the face of so many changes, so many tragedies and now a pandemic, the attachment to roots, to heritage, can perhaps give us some comfort and the illusion that we can go back on the roller coaster in the same way and with the same convictions as we had in the first experience and, in a symbolic way, "go against the course of history and suspend time" (Lévi-Strauss 1983, 9–10).

São Paulo, Brazil   Fabiana Lopes da Cunha
Ushuaia, Argentina   Jorge Rabassa

References

Certeau, Michel de, 2012. Culture in Plural. 7th Edition, Campinas, São Paulo: Papirus.

Eksteins, Modris. 1991. The Sacrament of Spring: the Great War and the birth of the modern era. Rio de Janeiro: Rocco.

Gonçalves, José Reginaldo Santos. 2015. The uneasiness in the patrimony: identity, time and destruction. In: Estudos Históricos. Rio de Janeiro, vol. 28, no. 55, pp. 211–228, January-June.

Lévi-Strauss, Claude. 1983. Un autre régard. L'Homme, Paris, 33, 126, pp. 9–10.

Sevcenko, Nicolau. 2001. A Corrida para o Século XXI: No Loop da Montanha-Russa. São Paulo: Companhia das Letras.

Acknowledgements

We would first like to thank all the Professors and researchers, who were invited to the event and who made it possible to debate and disseminate knowledge on the subject and also to publish this book and also the organizing and scientific commission of the event and all those who participated and contributed not only to the organization of our II International Symposium Heritage, but also the public who attended and participated in the debates, conferences, round tables, mini-courses and cultural activities.

An event of this size cannot happen without financial and institutional support. Therefore, our special thanks to:

the São Paulo State Research Support Foundation-FAPESP (process 2018/15765-6) for the financial support for the holding of the "II International Symposium on Heritage: Culture, Identities and Tourism";

the Paulista State University "Julio de Mesquita Filho"-UNESP/Campus of Ourinhos, mainly for the executive coordination and course coordination, for the financial aid and logistic and institutional support;

the Documentation and Memory Centre—CEDOM; CENPEA;

the Ourinhos City Hall;

the Municipal Secretariat of Culture of Ourinhos;

SISEM-São Paulo;

CADIC-CONICET, Ushuaia, Tierra del Fuego, Argentina;

CICOP-Brazil; Geocart; PIBID; NEPAM/Unicamp;

and the Heritage Research Group-CNPq.

We also want to thank those teachers, staff and students, who were very relevant to the success of the event, especially: Prof. Ph.D Luciene Cristina Risso (who helped us to think and organize the event), Associate Prof. Edson Luis Piroli (who greatly supported us in its management as executive coordinator of the campus), Profs. Ph.D Clerisnaldo Carvalho, Andrea Aparecida Zacharias, Prof. Master Carlos Henrique da Silva, Prof. Ph.D Paulo Fernando Cirino Mourão, Doctor Felipe Yera Barchi, Doctoral student Milena Santos Mayer and Master Rafaela Sales Goulart. These last three colleagues helped us a lot during the event, inserting information in the website, in the social networks and also in the organization of the Annals. Also, among the employees of the University and the students of the graduation section, our special thanks to Rodrigo Fantinatti Carvalho, Glaucia Marques da Rocha Coelho Garcia, Marcio Ribeiro Lopes, Reinaldo Bezerra Alves, Mariana Paula Umino, Leonardo Hiroshi Horie, Fábio Domingues, Alexandre Rogério Trindade, Paulo Alberto Augusto, Rafael A. Godoy da Rocha, Ana Luisa de Melo Antunes de Ávila, Edmilson Ferreira da Silva, Gabriela Botoni Vieira, Geyce Iris Goering Maia, Juliana Moreira, Jullio Cezar B. Gomes, Lucas Bitencourt, Maria Sarah Conca Parede, Marília Lopes, Nathalia Gomes, Vanessa Gomes, Weber Carvalho.

About This Book

This book is the result of the conferences and debates that took place at the **2nd International Symposium on Heritage: Culture, Identities and Tourism**, which took place from 22 to 26 April 2019. The event was promoted by the Universidade Estadual Paulista "Júlio de Mesquita Filho"—UNESP, Ourinhos, São Paulo, Brazil, and sought to bring together students, Professors, researchers, professionals in the field and the like, aiming to exchange experiences and strengthen the discussion of the importance of heritage through its multiple themes.

The symposium aimed to discuss themes in the area of Material and Intangible Cultural Heritage, seeking to contribute to the understanding, in a plural way, and appreciation of the scientific-academic and social universe of this theme. To this end, the event focused on multidisciplinary and interdisciplinary dialogue, showing the richness of the diversity of views and senses about the different forms of heritage and the importance of building the identity of communities and population through heritage in its most different expressions: such as festivals and tourist potentialities. For this, it was structured in two thematic sessions that contemplated more specific themes: Intangible Cultural Heritage and Material Cultural Heritage. These sessions have been divided into the following thematic axes: **Festivals, Heritage and Tourism**; **Cultural Landscape and Tourism**; **Railway Heritage and Museums** and **Archaeological Heritage and Tourism**. These themes were discussed in round tables, oral sessions, mini-courses, besides an opening conference and cultural activities. In this second edition, we seek to enrich the dialogue between the community of national and international public and private universities, with museums, with public institutions/organizations linked to the theme, representatives and professionals in education at all levels with interest and/or performance in the theme and members of society in general.

The book was organized with the same themes as the sessions that took place during the Symposium. Therefore, **Part I** of the book entitled *Festivities, Heritage and Tourism* has six chapters dealing with the theme in different places (London, Colombia, Portugal, Rio de Janeiro, São Paulo and Uruguaiana and their connections with Argentina and Uruguay) and about different musical genres and popular festivals such as carnival, the folia de reis, samba and fado. The Chapter "Atlantic Dialogues and Immaterial Heritage: The Party, Samba and Building a Community of Brazilians in London", entitled, is by Fabiana Lopes da Cunha. Its main objective is to discuss the samba of Rio de Janeiro, its cultural matrixes and its diffusion around the world through the instruments, practices and knowledge related to this musical genre and the biggest party connected to it: Carnival. In this work, the fruit of our previous studies on the subject and the post-doctorate studies held by Fabiana Lopes da Cunha at King's College London in 2017, we will seek to address the dialogue between Brazil, more specifically about the carnival of Rio de Janeiro and its dialogue with Notting Hill Carnival, through a case study: the Paraiso School of Samba.

The Chapter "Fado, Samba and the Interface between National Identity, Cultural Heritage and Tourism on both Sides of the Atlantic", by Felipe Yera Barchi and Fabiana Lopes da Cunha, aims to discuss how fado and samba developed as musical genres and became popular with the advent of radio and new media in the twentieth century. In addition, under a kind of state patronage, the two styles were elevated to symbols of Brazil and Portugal with the regimes of Getúlio Vargas and Antonio Salazar, respectively. The similarities between the two genres do not stop there, both were protected as cultural heritage and transformed into a tourist resource at the turn of the millennium. This intertwined trajectory holds many similarities on both sides of the Lusophone Atlantic Ocean and in this work our focus is on the contemporary processes of patrimonialization and consequent transformation into a touristic resource.

In "Importing the Spirit of Samba. Reflections on Intangible Heritage and the Reproduction of the Cultural Expression of Samba Outside of Brazil", by Yago Quiñones Triana analyses how Samba, as a form of cultural expression has been particularly developed in Colombia, exploring some facts related to the project of a group of artists who aimed to recreate the Brazilian Samba culture in Colombia, and the challenges they have encountered. This specific expression will be discussed and analysed through the lenses of some ideas of intangible cultural heritage. Specifically, it has been identified that there is an intention to reproduce something more than a musical rhythm or a dance, but there is a purpose to revive a concept that is difficult to define and that has to do with the life experience related to Samba. This idea of the "spirit" of Samba is what it is meant to be reproduced in Colombia; however, it is important to question if this is possible to achieve, and to identify the cultural barriers that may be found in the cited process. It is also relevant to conclude if the "spirit" of Samba is real or if it is an idealistic construction, and ultimately what is the relation of this case study with the reflections on intangible heritage.

The Chapter "Schools of Samba in the South Brazilian Border: Circuits and Translocal Exchanges in Carnival Cultures", by Ulisses Corrêa Duarte, is based on a multi-situated ethnography in the Pampas region, developed on the triple border in southern Brazil between Argentina and Uruguay, we will reflect on the carnivals of the schools of samba that occur beyond the center of the country. We will analyse the social–cultural importance of the carnival in Uruguaiana (Brazil) and the consequent impacts of the festival on local politics, the tourist sector and the economic dimension of the event in the region. The extended carnival calendar of schools of samba in the Pampas promotes a ritual time of the festival that allows for extensive exchanges and trans-local negotiations between the carnivals on the border and the carnivals of schools of samba in Rio de Janeiro. The carnival circuits in the Pampas allow us to think about hybridism, globalization, and the cultural identities produced by popular cultures on the margins and beyond the borders.

Continuing the discussion on samba schools, our Chapter "Igor Sorriso: A Narrative About Experiences and Evolution of Life Through Carnival" is about an autobiographical testimonial by the great samba singer Igor Sorriso, which was transcribed by Rafaela Sales Goulart and revised by Fabiana Lopes da Cunha. This chapter tells us about his life, his discovery as a musician and lover of samba and how his experiences were in various places in Brazil and, especially, as a samba interpreter in the great samba schools of Rio de Janeiro, such as Vila Isabel and, of São Paulo, in Mocidade Alegre. It is important to emphasize that, among his professional autobiography, Igor reveals how much he learned from his career related to Carnival: his perception about caring for his own voice, his contacts and identifications that made him move to São Paulo and, finally, the recognition that the association/pavilion is the main flag of the Carnival and, therefore, its main patrimony.

And finally, our last chapter of this session is "Religious Celebrations and its Safeguard Public Policies in Brazil: Directions" by Rafaela Sales Goulart and Fabiana Lopes da Cunha, is based on the explanation of the historical trajectory of the heritage concept and, more specifically, its appropriation as a political object of identity projections, this manuscript will reflect on the unfolding of cultural heritage in Brazil, emphasizing the field of intangible heritage. In this field, in turn, it is found the category "celebrations", created as a result of the instrumentalization process of public safeguard policies for such cultural goods (Decree No. 3,551, of August 4th, 2000). In this context, we aim to discuss the main limits and possibilities that encompass studies, registration processes and social actions that favour the safeguard and continuity of these celebrations, especially the religious ones. The directions have been chosen and will be shared not only because of our research experience with groups of folia de reis from the interior of the state of São Paulo, but also because of the challenges inherent in instrumentalizing public policies for such celebrations.

Our **Part II** of the book, *Cultural Landscape and Tourism*, has two chapters dealing with the subject. The Chapter "Perception of the Cultural Landscape in Historical Centers" of the book, by Rosio Fernández Baca Salcedo, deals with how the cultural landscape has an impact on our senses, on our activities, interfering in decision-making, judgment and values and in communication with the urban space. The interrelated dimensions, necessary to understand the perception of the cultural landscape, are the physical, social and symbolic dimensions. In this article, we address the perception of the landscape in historical centres in the physical (categories: group/dispersion, integrate/segregate, attract/repel, open/close, walk, sit and stand) and social (categories: necessary, optional and social) dimensions.

The Chapter "Landscape, Heritage and Tourism: Study in the Historic Center of Seville—Spain" by Luciene Cristina Risso, approaches the thematic of heritage landscapes and growing world tourism practices, taking as a case study, the historic centre of Seville, with special emphasis on the Triana neighbourhood. The main objective was to identify the main cultural heritage landscapes of the historical center of Seville and the pressures arising from the touristic industry. For this, the methodology involved readings, data collection and geographical interpretations. In the survey of the primary data on the cultural heritage of Triana, there was an ethnographic description and participative observation of the Triana festivities (Holy Week, Santa Ana festivity, Flamenco show) and, for the perceptual dimension, the leaders of the Northern Triana Neighborhood Association were interviewed. In the collection of the secondary data, the identification of the cultural heritage or assets of cultural interest (i.e. "Bien de Interés Cultural"—BIC) was used through the general catalog of the Andalusian historic heritage of the Council of Culture of the Junta of Andalucía (until 2016) and the tourism data were collected in the tourism council of Seville, referring to the years of 2017 and 2018. As a result, it is noted that most part of the cultural heritage of the historic centre fit in the typology of monuments legitimized by elites from different historical periods, which receive a massive amount of tourists and, in Triana, the ethnological heritages were a landscape potential, added to its history and symbolism, equally threatened.

Our **Part III** of the book, entitled, ***Rail Heritage and Museums***, has four chapters. The Chapter "Museums, Railway Memories and Cultural Landscapes", entitled, by Davidson Panis Kaseker, connects the third and fourth sessions of this book. This chapter talks about the state of São Paulo, despite its more than five thousand kilometres of railroads and hundreds of stations, warehouses and residential complexes destined for railway workers, only 12 items are listed by the Instituto de Preservação do Patrimônio Histórico e Artístico Nacional (IPHAN) (i.e., Institute of National Historical and Artistic Heritage). In the list of the Council for the Defense of the Historical, Archaeological, Artistic and Tourist Heritage of the State of São Paulo (CONDEPHAAT), there are 32 railway buildings listed. The scarce number of railroad properties under the protection of the State's preservation agencies shows the selective criteria adopted by the official policies of historical heritage preservation. There are no official statistics, but it is known that many cities claim the historical value of their respective railroad heritage, taking into account the cultural relevance, usually associated with the origin of their own urban formations.

Since the turn of the twenty-first century, the Brazilian and São Paulo railroad system has been in the headlines of newspapers and TV stations spreading images of abandoned and depleted stations, rolling stock depreciated by the weather and disuse, passenger cars, locomotives and wagons rotting in urban yards. Theft of rails and poles became a routine agenda until the plundering was completed. The demand for the preservation and musealization of this type of heritage, stimulated by the bonds of affection in the imagination of a large part of the population that cultivates the railway memory, intensified in another way.

The objective of this text is to point out guidelines for the musealization of open-air railway heritage, taking into account the methodological principles of museology and the distinct potentialities of cultural appropriation of these tangible and intangible assets by the populations of its surroundings with a view to the recognition of their cultural value and the regeneration of urban texture and social tessitura.

Our Chapter "Sorocabana Station in Bauru (São Paulo, Brazil): Diagnosis for Restoration" talks about a specific case study of a city in the interior of São Paulo State, Bauru, by Nilson Ghirardello, discusses the railway network in Bauru, a city created at the end of the nineteenth century as part of the territorial occupation process of western São Paulo, where coffee plantations for export, followed by the railroad, played a key role. Three railroad companies were installed in the city: Companhia Sorocabana (1905), Northwest of Brazil (NOB) (1905) and Companhia Paulista (1910). The most expressive for the urban development of the city was the Northwest of Brazil (NOB) railway company, due to its headquarters in Bauru, its size and the significant strategic and pioneering role towards the west of the country. However, the NOB company only established its headquarters in the city due to the pre-existence of the Sorocabana Company station. The historical importance of this small construction, still existing, is its size; in 2015, a project for its restoration was proposed. There are two main objectives of this work: to draw a history about the installation of the railroads in Bauru and to present the diagnosis of conservation of the building of the old Sorocabana station, used for the elaboration of the restoration project, but unfortunately, not yet executed.

The Chapter "Heritage and Museums: The Cultural Significance of the University", by Paulo Henrique Martinez, is a reflection on the impacts of the social and cultural changes of the second half of the twentieth century on the daily work with heritage and museums at the Brazilian university. The proposition of the United Nations social agenda from the 1990s up to 2030 created possibilities for the development of individual, institutional and social capacities in teaching, research, university extension and cultural actions from heritage and university museums.

The Chapter "Tropas and Tropeiros in Southern Brazil: History, Memory and Heritage" and the last of that session, by Milena Santos Mayer and Fabiana Lopes da Cunha, is the result of a dialogue between the authors based on a doctoral research in History, still in progress, which has as its central object the trajectory of a museological institution called *Museu do Tropeiro* (Museum of Tropeiro). This museum is located in the interior of Brazil, in the city of Castro (Paraná State) and it is dedicated to the preservation and dissemination of the history and memory of the mule trade in the southern region and its social and cultural implications in the municipality and the region. This was an activity that began in the colonial period; when the need arose to transport cargo and beef animals throughout the Brazilian territory. This practice can be evaluated as a global phenomenon, since the use of animals was for a long time the main means of transportation for humanity. However, in southern Brazil, this activity has developed with the peculiarity of the significant commercialization of mules. These animals were transported from the region of the Brazilian pampas to the city of Sorocaba, located today in the state of São Paulo, and then sold to be used elsewhere in the country as means of transportation for people and goods. In this way, long roads were built that made possible the integration of a part of the Brazilian territory that was far from the relatively better known coast. The *tropeiros* (or muleteers), men who drove and traded these animals, had the need to stay overnight in certain places for their own rest and for the reestablishment of the *tropas* (or trains). One of the main stopping points was the Campos Gerais region in the current state of Paraná, propitious for its field vegetation, the region has as its oldest administrative organization the current municipality of Castro. In 1977, when the city conquered its public museum, it emerged as a thematic museum, the first in the country dedicated to the history of *tropeira* activity. However, researching the institution and the subject in question, it can be seen that later on other museums, memorials and collections are established in several places in the country. Therefore, this article deals with Brazilian historiography in relation to the subject and the construction of places of memory of the *tropeiro* in Brazil and its implications in relation to the resonances of cultural heritage taking Castro's museum as a reference.

Part IV of the book, ***Archaeological Heritage and Tourism***, has three chapters with important reflections on the subject.

The Chapter "Illicit Trafficking in Cultural Assets: A Genealogy of the Concept and Actions in Contemporary Brazil", by Rodrigo Christofoletti, which brings a relevant discussion on how a significant part of cultural heritage has been taken by a multimillionaire trafficking system and documents from international organizations estimate that art and heritage trafficking tops the list of the biggest illicit actions in the world, just behind drugs, weapons and human trafficking, which created a well-articulated systemic grid that indicates an exponential growth trafficking pattern. Such a mesh feeds a fairly complex system. Some examples of this mesh are private collectors, museums, monuments, religious sites, archaeological/paleontological sites and other private preservation spaces; illegal excavations (including underwater excavations); theft of artifacts and works of art during armed conflicts and military occupations; illegal downloading of miscellaneous properties; production, exchange and use of forged documentation; even trafficking of cultural goods, authentic or counterfeit. This whole range of actions linked to trafficking has been fought in recent decades as the life of this set of goods is increasingly in danger. This article aims to address this type of illicit trafficking, suggesting that an international art trafficking route has in Brazil one of the least studied but no less important capillary points.

The Chapter "Jesuit-Guarani Missions: UNESCO World Heritage Site in Brazil", by Tobias Vilhena de Moraes and Pedro Paulo Abreu Funari, aims at presenting some aspects of the concept of the archaeological preservation of the remains of material culture and the role of the archaeologists during site management planning at the archaeological Jesuit-Guarani Missions, a UNESCO World Heritage site in Brazil. Located in the southern part of the South American continent (in the state of Rio Grande do Sul), the Missions were excavated for the first time and conserved from the beginning of the 1980s. Since then, many conservation projects have been carried out at the site in an effort to obtain a more accurate method to site management planning. In this chapter, we present the recent results of site management planning at the Jesuit-Guarani Missions World Heritage site and its implications to heritage legislation in Brazil.

And finally, to conclude our last session and the book, the Chapter ""Invisible Heritage": New Technologies and the History of Antarctica's Sealers Groups", by Andrés Zarankin and Fernanda Codevilla Soares, discusses alternative forms of heritage construction and preservation related to the human occupation of Antarctica by subaltern groups and its invisibilities in the official discourses on the colonization of the continent. Our proposal associates public, digital and sensorial archaeology approaches, highlighting a more pluralistic and democratic narrative about the southernmost past. For this purpose, we used tools such as new technologies applied to archaeological research (3D laser scan, object scanner, 3D printers, drones and others). Besides, through itinerant sensory exhibits (which simulates the Antarctic environment within an inflatable dome), we seek to narrow communication channels between the archaeological and non-archaeological public. Also, we encouraged the construction of multivocal narratives on the human occupation of Antarctica.

Contents

Rail Heritage and Museums

Archaeological Heritage and Tourism

Festivities, Heritage and Tourism

Atlantic Dialogues and Immaterial Heritage: The Party, Samba and Building a Community of Brazilians in London

Fabiana Lopes da Cunha

Abstract The main objective of our research is to discuss Rio de Janeiro's samba, its cultural matrixes and its diffusion around the world through the instruments, practices and knowledge related to this musical genre and the biggest party linked to it: Carnival. In this work, the fruit of our previous studies on the subject and the post-doctorate held at King's College London in 2017, we will seek to address the dialogue between Brazil, more specifically about the carnival of Rio de Janeiro and its dialogue with Notting Hill Carnival, through a case study: the *Paraiso School of Samba*. (*Author's note*: We have chosen here to insert the name of the Samba School as it is widespread in London. The literal translation into English would be: *Paradise School of Samba*)

Keywords Atlantic Dialogues · Immaterial Heritage · Samba · Notting Hill Carnival · Brazilian Culture

1 In the World of *Macunaíma*

> One must begin to loose one's memory, even fragments of it, to realize that it is this memory that makes one's whole life. A life without memory would not be a life, just as an intelligence without the possibility of expressing itself would not be an intelligence. Our memory is our coherence, our reason, our action, our feeling. Without it, we are nothing. (Buñuel 1982, p. 11)

Buñuel's quote, with which I begin the text, denotes the relevance of our memories as a valuable individual heritage. Without them, we loose our identity, our stories, our experiences. The discussion, the inventory and the record of heritage of immaterial or intangible nature are also very much linked to the importance of these individual and collective memories, the daily cultural practices and their preservation through the record and practices of safeguarding these assets.

F. L. da Cunha (✉)
Paulista State University "Julio de Mesquita Filho"—UNESP, São Paulo, Brazil
e-mail: fabiana.cunha@unesp.br

© The Author(s), under exclusive license to Springer Nature Switzerland AG 2021
F. Lopes da Cunha and J. Rabassa (eds.), *Festivals and Heritage in Latin America*,
The Latin American Studies Book Series,
https://doi.org/10.1007/978-3-030-67985-9_1

3

Since the implementation of these instruments to safeguard the intangible heritage in Brazil, with the four record books (Knowledges, Celebrations, Forms of Expression and Places), many inventories and records have been made in different regions of the country, and can be consulted on the IPHAN (*Author's note*: Institute of National Historical and Artistic Heritage—Brazil. Available at http://portal.iphan.gov.br/pag ina/detalhes/234).

Most of these 47 records (made until the year 2018), spread throughout the Brazilian territory, have a direct relationship with the music and the orality (that is, the quality of being verbally communicated) of the groups and communities involved. This denotes the relevance of music within the policy of safeguarding the intangible heritage in our country and the connection with the origin of IPHAN, with its creator, Mário de Andrade.

In his work, *Macunaíma*, Mário de Andrade makes a long aesthetic reflection that has, according to Souza, two constant points of reference: "the analysis of the musical phenomenon and the creative process of the people". (Souza 1979, p. 11) The work, written in 6 days, in December 1926 and expanded in January 1927 and published in 1928, was elaborated during the author's vacation and built from elements of "oral or written tradition, popular or erudite, European or Brazilian" (Andrade 1982, p. 76) Souza's analysis of the modernist's work denotes his conviction that Mário de Andrade "intentionally and critically transposed into literature the conflict he had observed so acutely in music between the European tradition inherited from Portugal and local, popular, indigenous or Africanism manifestations" (Souza 1979, p. 74). In a way, this discussion had already been prefigured in part by Machado de Assis's relentless imagination in his tale "A Famous Man" which tells the story of the composer known as Pestana. Sevcenko pointed out in his text that the composer was a singular creature. He was the "most famous and celebrated popular composer of his time, a period that Machado framed between the 1870s and 1880s.

> Any of his compositions, as soon as he arrives at the music houses, reveals an instant and overwhelming success. They are performed all over the city, in the domestic pianos, in the bands that liven up the parties, in the chanterolas entertained in the bars and in the offices, in the whistles of the passers-by. Pestana is undoubtedly the most loved and praised person in Rio de Janeiro. An authentic famous man.
>
> The only problem is that he hates that reputation. His true and obsessive ambition in life is to accomplish at least one page of classical music, one only. The walls of his house are covered with images of the great immortal masters, Bach, Mozart, Beethoven, Schumann, Chopin… But when he sits at the piano and plays, what comes to his hands, inevitably, is the most sapeca and delicious polka that Rio de Janeiro brejeiro has ever damaged. There are only two things he hates more than naughty popular music: the sensual rhythm and the dancing impulse. Exactly his forte. Pestana feels repugnance for what he is and for what he does. (Sevcenko 2004, pp. 13–14).

His nature divided between his father, who some said was a priest, who taught him Latin and music, with a European profile, and his mother, who by physical description, would be brunette, of Brazilian origin.

> Educated and haunted by the disciplinary patriarchal figure, Pestana's moral identity is forged in Old World values. Orphaned and stigmatized by the dubiousness of birth, the feeling of

guilt, shame, local culture and subconscious impulses. It is the feminine side that he denies, the mother and, by extension, the Brazilian culture. But it is precisely this self that always wins, to the general joy of the people. (Sevcenko 2004, p. 14)

As well as the divided nature of Pestana, the material that served Mário de Andrade in the elaboration of the Macunaíma narrative, we see an ethnic mixture of popular music evidenced in the variety of elements coming from the most diverse sources, from indigenous traces taken from distinct works of Koch-Grünberg to Capistrano de Abreu plus "ceremonies of African origin, evocations and songs of Iberian origin, Portuguese traditions, tales already typically Brazilian, etc." (Souza 1979, p. 16).

And this mixture, this hybridity is not conclusive, it is ambivalent and in several moments the author described the behaviour of the character: "always torn between the two fidelities, to Brazil and to Europe"(Souza 1979, p. 96).

"I'm a tupi tangendo (*Author's note*: Meaning similar to playing) **a lute"**...

This fracture and ambiguity is also present in this famous phrase by Mário de Andrade, which Gruzinski used in his reflections on the History of Culture during his stay in Algodoal, where chords from a "primitive" harp rhythm the evolution of the boys who danced capoeira on the beach. For the scholar, "the mixture would invariably be under the sign of ambiguity and ambivalence. Such would be the curses that would hover over the mixed worlds" (Gruzinski 2001, p. 26). For him, the "archetype of the Brazilian and the Latin American, divided between antagonistic options—Brazil or Europe-, oscillating between cultures, but belonging simultaneously to all is demonstrated in *Macunaíma* from beginning to end, as a divided being". (Gruzinski 2001, p. 26).

Macunaíma's outcome illustrates the impossibility of escaping from the contradictions and dilemmas of double bonding. We would particularly say that they are not just two worlds, two bonds...there are several! In search of a wife, the hero hesitates between the two worlds: he chooses, one after the other, a Portuguese one, and then an Indian woman, Miss Sancha. But his choice solves nothing. If, at last, he is attracted to the native, it is because the girl's mother, *Vei*, the Sun, gives him the appearance of a European. *Macunaíma* fell for the trap that *Vei* had set her up. The hero's mistake expresses the complexity of the situations that arise from the confrontation of worlds.

But, looking closer, *Macunaíma*'s contradictory decisions do not cancel each other out. The two successive choices, "the two sequences do not cease to form a perfectly organic whole within the structure of the story. In *Macunaíma*, the antagonistic elements present themselves as "the two faces of the same coin". It is impossible, therefore, to dissociate them. Like the inhabitants of Algodoal, Macunaíma fully feels the attraction of the Western universe. For it is part of that universe as much as they are (Gruzinski 2001, p. 27).

But, if such a fracture is felt and thought about inside Brazil, what would this relationship be like outside the country? How do Brazilians living abroad live, feel, spread such a culture?

Some questions from Stuart Hall also stimulated us to think about the experience of Brazilians living abroad, and more specifically, in London.

What does the experience of the diaspora cause to our models of identity and cultural heritage? More specifically, the immaterial or intangible?

How can we conceive or imagine identity, difference and belonging after the diaspora? How do we translate and are we translated in another country, in this case, in England, more specifically, in London?

Since "cultural identity", according to Hall, carries with it so many traces of essential unity, primordial uniqueness, indivisibility and sameness, how should we "think" of the identities inscribed in power relations, built by difference and disjuncture? (Hall 2009, p. 28).

Just like the Caribbean people, treated by Hall, our history is also marked by ruptures: by conquest, expropriation, slavery, the system of ingenuity and long colonial dependence. Our people also have roots in different places in the world, which results in a "creolization" of the "transcultural" type: it is a "process of 'contact zone' that invokes the spatial and temporal co-presence of subjects previously isolated by geographical and historical disjunctures [...] whose trajectories now cross, and such logic and perspective is as interested in how the colonized produces the colonizer as vice versa". (Hall 2009, pp. 28–30).

The binary concept of difference on which the closed concept of diaspora is based is guided, according to Hall, by the "construction of a border of exclusion and depends on the construction of an "Other" and a rigid opposition between the inside and the outside". However, the syncretized configurations of the Caribbean cultural identity, like the Brazilian one, require the *Derridian* (that is, relating to the French philosopher and critic Jacques Derrida and his ideas) notion of *différance*:

> a difference that does not work through binarisms, veiled borders that do not finally separate, but are also *places of passage*, and meanings that are positional and relational, always sliding along a spectrum without beginning or end. Difference, we know, is essential to meaning, and meaning is crucial to culture. But in a profoundly counterintuitive movement, modern post Saussarian (i.e., characteristic of the theories of Ferdinand de Saussure, especially the view that a language consists of a network of interrelated elements in contrast) linguistics insists that meaning cannot be fixed definitively. There is always the "inevitable slide of meaning into the open semiose (that is, the process of production of linguistic signification) of a culture, while what seems fixed remains dialogically re-appropriated. The fantasy of a final meaning continued haunted by "lack" or "excess", but is never apprehensible in the fullness of its presence to itself. (Hall 2009, p. 33)

And it complements with arguments from Bakhtin and Volochinov:

> The social plurivalence of the ideological sign is a trait of the greatest importance (...) in fact, it is this intertwining of the value indices that makes the sign alive and mobile, capable of evolving. The sign, if subtracted from the tensions of social struggle (...) will infallibly weaken, degenerate into allegory, become the object of study of philologists (Hall 2009, p. 33).

In this conception, the binary poles of "meaning" and "non-sense" are constantly ruined by the more open and fluid process of "making sense in translation".

Kobena Mercer described this cultural logic as "dysphoric aesthetics". In a whole range of cultural forms, there is a powerful syncretic dynamics that critically appropriates elements of the master codes of dominant cultures and "creolizes" them, disarticulating certain signs and otherwise re-articulating their symbolic meaning. (Hall 2009, p. 34). Using Hall's reading of Caribbean culture, we understand that Brazilian culture is also essentially driven by a "diasporic" aesthetic because, in anthropological terms, its cultures are hopelessly "impure".

And, according to Arantes (2004)

> These local realities that cross ethnic and territorial borders generate inter-textualities, which must absorb characteristics of the same code.
>
> Globalised cultural production thus provides for the rooting, at local level, of global senses of place, senses that dialogue, dislocate and interact with representations of identity, memory and tradition, and with the practices associated with them. (Arantes 2004, p. 13)

Despite all these difficulties pointed out by these authors, as well as the one discussed by Arantes, samba and carnival, at least in London, where we conducted our research, the rite and cultural manifestations have not lost their sense of tradition, but have changed and this we were able to notice in our research. But we understand that they go through a process of reinvention of their own ethnicity (and, for our understanding, of different nationalities), as Gilroy discusses in his "Black Atlantic", a term that describes the fusion of black cultures with other cultures from around the Atlantic Ocean.

Gilroy discusses in his work the historical and political significance of black cultures and the question of their identity and not identity in the United Kingdom. For him, as far as the question of music in what he defines as the "black Atlantic world" is concerned, these are "primary expressions of the cultural distinction that this population captured and adapted to new circumstances". (Gilroy 2001, p. 173) For him,

> If these populations are somehow unified, it is more by the experience of migration than by the memory of slavery and the waste of the plantation society. Until recently, this same novelty and lack of roots in the "native" cultures of the urban [inner cities] of the United Kingdom conditioned the formation of racial subcultures that were greatly influenced by a kind of "raw material" from the Caribbean and Black America. (Gilroy 2001, p. 173)

Obviously, there are rules that have to be discussed with the local government and the Caribbean community, which created the Notting Hill Carnival. "New rules and criteria" are negotiated between the "natives" and the "gringos". These depend on the space where the samba and carnival are being executed. In the specific case of London, the Notting Hill Carnival takes place in the neighbourhood that gave it its name and began with a political and cultural performance and affirmation of the Caribbean community that migrated to the UK in the post-war period. The two samba schools that participate in the London Carnival, the *London School of Samba* and the *Paraiso School of Samba*, try to recreate in the streets on the London Carnival circuit, which takes place in spaces built for the parade in Brazil, such as the *sambodromos* (*Author's note*: Spaces built in some Brazilian states for the Samba Schools to parade with their great Allegorical Cars and with their thousands of members).

Of course, this is a very complex activity and has a distinct result from what happens during carnival in Brazil, especially in places like Rio de Janeiro and São Paulo. But the effort that the components from other countries, who are part of these schools, make to understand samba and carnival is great. Many Europeans and Caribbean people who participate in the *Paraiso School Of Samba* travel to Rio de Janeiro (some to São Paulo) in order to learn and experience samba and carnival, mainly by observing and playing on the drums of the samba schools. They learn Portuguese through and because of the music, the rhythm, the party. Here they dialogue with the communities. In London, it is the Brazilians who join with the English, French, Irish and Caribbean people to make the musical and festive manifestation possible. However, it is evident when talking to the "foreigners" who are lovers of this musical genre (the samba) and this great festival, carnival, the recognition of the cultural value of the knowledge and practice of the drummers and members of the samba community in Brazil.

Thus, from the internal point of view of culture and social experience, as Arantes (2004) states, "product and process are inseparable".

> They also harbour the feelings, memories and senses that are formed in the social relationships involved in production, and thus, the work feeds life and human relationships. (Arantes 2004, p. 17).

And this work, produced by generations of practitioners

> of a particular art or craft is something more general than each piece produced or executed, than each celebration held. It is knowledge; it is technology; it is verbal, graphic, scenic, choreographic and musical language; they are collective and diffuse world views. But, on the other hand, one finds in each work or in the memory one has of it, the testimony of what someone is capable of doing. The product made encloses the individual authorship and the collective making, the capacity to repeat a gesture and to modify it, keeping alive - but never identical - the tradition, since in the phrases said, the language is perpetuated and constantly renewed. (Arantes 2004, p. 17)

These strategies and negotiations will enable samba, a musical rhythm linked to candomblé (an Afro-Brazilian religion), circuses, religious parties and hills of Rio de Janeiro, in the first decades of the twentieth century, to become the musical genre of our biggest party, carnival, and of persecuted and marginal music, a symbol of Brazilianness (Cunha 2004).

Samba has become one of the most relevant expressions of our intangible or intangible heritage and has extrapolated its communities, and today there are several groups linked to samba and carnival in various places around the world (European countries, Japan, China, Canada, among others). Our post-doctorate at King's College London had as its main objective the analysis and relevance of this musical genre in London and its diffusion through the biggest street party in Europe: the Notting Hill Carnival. Through interviews, document analysis, newspapers, blogs, social networking, websites, listening and analysis of sambas-story we sought to understand this phenomenon of Brazilian culture in London through the reconstruction of the history of a samba school, the Paraiso School of Samba, its social actors, its community, its events to answer the question of how samba and carnival connect people and the two sides of the Atlantic Ocean: Brazil and the UK. With our research, we seek to understand through the phenomenon of multiculturalism these connections, these bonds of identity and try to answer what types of "community" these individuals form, what their relationship with British society, what strategies for their integration into that society, as well as understand the "links of continuity with their places of origin". And also, what other links are built through music, dance and partying. More specifically, at the Notting Hill Carnival. By experiencing this experience, we agree with Arantes (2004) that the circulation of cultural goods

> are among the main ingredients of the changes that occur in lifestyles and the formation of symbolic boundaries across the planet. But it is always useful to insist that far from simply creating homogeneity, the global market stimulates the generation and circulation of all kinds of resources capable of producing senses of place and difference. (Arantes 2004, p. 13).

2   Brazilian Diaspora

By doing interviews with several Brazilians who had migrated to London, at different ages, it was possible to assess that each one had a reason or expectation when going to London, but none of them, or at least the majority of them, thought, when leaving Brazil, of living the rest of their lives in the capital of England. However, they all felt a very strong empathy with the city, almost instantly, and ended up choosing to live in the English capital. London seems to be really welcoming for many of the immigrants. At least that is what I noticed in the various interviews I did with the Brazilians who decided to stay in the capital of England. It is an organized city, with lots of opportunities and multiculturalism. You can see it all the time through the streets of the city. And, this has also made it possible to organize and demonstrate Brazilian culture on the streets of the city. Such events and Brazilian presence in the cultural scene of the city is related to the increase of Brazilian immigrants in the UK. If Brazilians have been coming here since the last decades of the twentieth century, it is possible to see, according to studies (Evans and Souza 2015, p. 7), that not only has the number of Brazilians in the city increased, but their profile has also changed since the 2000s.

In the Brazilian academic environment, in the consulates and NGOs that help Brazilians abroad, it is a consensus that in the past three decades many Brazilians have come to live abroad. According to Jan Brzozowski (2012), Beatriz Padilla (2006) and Olivia Sheringham (2013) "[…]Brazil can be glimpsed today as a diasporic nation, whose population is present in different countries of the European Union and North America" (Evans and Souza 2015, p. 7). It is possible to see this by the numbers in the table below, although inaccurate, because many Brazilians live illegally abroad:

Analysing the Table 1 we can find out that the UK is the third most sought-after destination by Brazilians for residence in Europe, at least until 2012, and this extends to the study cited from 2015.

According to some studies on the subject (Evans et al. 2007, Vertovec 2007), the Brazilian presence in the UK became stronger in the early 2000s.

It is possible to see this presence of "Brazilians" in London from the observation of the occupation of certain spaces and the emergence of institutions to help these immigrants:

Table 1 Main European countries with a presence of Brazilians (Evans and Souza 2015, p. 7)

Country	Brazilians
Portugal	140,426
Spain	128,238
United Kingdom	118,000
Germany	95,160
Italy	67,000
France	44,622
Sweden	44,089
Belgium	43,000

Source Ministry of Foreign Affairs (2012)

The presence of the food trade (which includes restaurants, grocery stores, butchers and cafés; see Aguiar 2009, Brightwell 2010), the beauty industry (which includes product shops, manicures, massage therapists, beauticians and hairdressers; see Figueiredo 2010) and the culture industry (see Frangella 2010) is visible in several London neighbourhoods. It is also easily noticed the presence of organizations that aim to provide assistance to this group of immigrants. There are several religious organizations with links to Brazil (see Sheringham 2013; Souza 2016), as well as Brazilian educational institutions (see Souza and Gomes 2016) and those that provide assistance with sexual health and immigration issues (Evans and Souza 2015, p. 6).

Of this phenomenon of the emergence of institutions and manifestations of Brazilians in London, the one that interests us most in this work is the cultural industry, especially the one related to samba and carnival. According to Frangella (2010), "London has received a growing flow of Brazilian cultural trends, styles and products, through two movements that run in parallel […]".

On the one hand, there is a significant diversity in the production and innovation of Brazilian culture, promoted in part by the growing emphasis by and for Londoners on fashion images (trendy) of Brazil as a cultural product. On the other hand, there is a significant increase in Brazilian immigrants in the UK in this decade, and consequently a heterogeneity in the profiles of immigrants and the dynamics of commercial and cultural experiences that come with them. (Frangella 2010, p. 33).

Such presence is very visible in some neighbourhoods, establishments and in the distribution of newspapers and magazines aimed at the Brazilian community. In Queensway, where we lived for 6 months, for example, there is a Brazilian market and restaurant, hairdressers, a butcher's shop that shares a partnership between products from Brazil and Argentina, and there was also a Brazilian rodizio (*Author's note*: Grilled meat offered in "rodizio" style (which ended its activities in 2019), owned by Italians, but which had a team of Brazilians for the care and preparation of food and meat.

At the *Brazil-London Foundation*, the newly arrived Brazilian can have access to important information about the city and also help and guidance on visas, accommodation and jobs, for example. There is also a guide with Brazilian restaurants in the capital of England. In some of these establishments it is possible not only to taste the flavours of Brazilian cuisine, but also to participate in events and listen to Brazilian music. But, there are several other places that are not on the site that have events related to Brazilian art and culture. On Mondays, there is a famous samba circle at Barrio East, in Shoreditch, with the group *Quintal do Samba*. There was also a bar, *Guanabara*, which had several events and presentations about Brazilian culture and which closed its doors. Today it is called *Tropicana* and, despite its name and activities, it's not owned by Brazilians. The *Paraiso School of Samba* performed there many times. In 2017, the *Piano Works* opened its doors to important events of the samba school, such as the Queen and Princess of Drums contest and also to the release of the samba-plot for the Notting Hill Carnival. *Made In Brazil Boteco* has a Brazilian DJ linked to the samba school, Fred Salgueiro, who promotes on one day of the week, the day of the school's rehearsal at Chalk Farm, the possibility for members of the association to go to the establishment, pay less what they consume and listen to good Brazilian music, played by the samba school's friend and trustee, Fred. The owner has partnerships with other institutions and personalities linked to Brazilian culture: a bilingual newspaper, the *Brasil Observer* (in Portuguese and English, which unfortunately closed its activities in October 2017) and also promotes projects and partnerships with Brazilian artists. Once a month, the *Brasil Observer* launched its monthly volume in the bar and promoted some artistic activity along with the event, in order to attract an even larger audience for the occasion: a musician, a theater artist, a dance performance. It was also possible, that day, to find several Brazilians who sought to promote events and links between the Brazilian community and London. On the day we participated in the launch of the publication, in August 2017, we met not only the owners and founders of the newspaper, a very interesting bilingual publication of excellent quality, but also a producer and promoter of events related to Brazilian art and culture and owner of *Braziliarty* that invited us to another event, focused on cuisine from several countries, denoting with the promotion of this activity the cultural dialogue between Brazil, Latin American countries and London: the Food Fest. We also had the opportunity to watch, in this same place, the show of a Brazilian artist who was on tour in Europe. (*Author's note*: Information from 2017)

But, going back to the question that motivated us to the above exhibition: "what is the role of migration in the circulation of these cultural goods? "This is the question that Frangella (2010) seeks to answer in her work through data collected for 4 years, with observation participating in cultural meetings of Brazilians and also through observations of public spaces. We also use these mechanisms for our work, associating it with interviews, document analysis, newspapers, blogs, social network, websites, listening and analysis of sambas plot in order to understand this phenomenon of Brazilian culture in London, seeking through the reconstruction of the history of a samba school, the *Paraiso School of Samba*, its social actors, its community and its events to answer the question of how samba and carnival connect people and the two sides of the Atlantic: Brazil and the UK. Our main aim with this work was to understand through the phenomenon of multiculturalism from the point of view of Stuart Hall (2009), these connections, these bonds of identity and to seek to answer what types of "community" these individuals form, what their relationship with British society is, what the strategies for their integration into that society are, as well as to understand the "links of continuity with their places of origin" (Hall 2009, p. 62). And also, what other links are built through music, dance and partying. More specifically, at the Notting Hill Carnival.

3   Brazil at the Notting Hill Carnival

London is the place for me,

London this lovely city

You can go France or America,

India, Asia or Australia,

But you must come back, to London city

We believe me I am speaking broad-mindedly,

I am glad to know my mother country.

I've been travelling to countries years ago,

But this is the place I wantin' to know.

London is the place for me.

To live in London you really comfortable

Because the English people are very much sociable.

They take you here and they take you there,

And they make you feel like a millionaire.

So London, that's the place for me.

(*Author's note*: Lord Kitchener (Alwyn Robert), *London is the Place for Me* (London, Honest Jon's records Ltd, 2002)- first recorded on newsreel in 1948. Apud. TOMPSETT, Adela Ruth. "London Is the Place for Me": Performance and Identity in Notting Hill Carnival". *Estudos de História do Teatro*; **Grand Forks, ND** 25 (2005): 43–60. p. 43)

When I read the lyrics of that calypso, first sung by Lord Kitchener to a reporter when he left the first ship that brought the West Indians to Great Britain, I could not help identifying with the part of the lyrics. Of course, we will see later that this was not exactly the scenario that the Caribbean people would experience in that context, in the capital of England.

But I want to point out that when I arrived in London, I felt as if I belonged here. I felt at home. I have visited other countries, but I have never felt so welcome. The city welcomed me. Institutions, communities and groups. In the streets of Bayswater where I wandered, I made friends in grocery stores and pubs. When I passed *Manoush*, *Prince Alfred* or the Arabian grocery store, as I called it, I was greeting people: "*How are you?*" And I always heard a "*good*" followed by a smile. There was no way, a quick conversation started.

For me London is colorful, despite the weather and the grey days. I arrived in mid-March, but a few weeks later the flowers were already blooming announcing Spring. The colors were not only on the flowers of the trees, gardens and vases in the houses and shops, but also on the clothes and the different ethnic groups and accents of different nationalities, which it was possible to glimpse and hear at every moment in the streets. The sounds could also come from different instruments that sang melodies with different rhythms. The bassist of the Queensway subway, the violin girl in front of the *Manoush*, or the rhythm of the atabaques of the young man who insisted on dancing and playing at the same time, in front of *Prince Alfred*.

But back to 1948 and *Mr. Calypsonian...*

The *West Indians* did not get the reception they had hoped for and sung the song. Hostility soon began and many Caribbean immigrants who came to London during this period lived in difficult and discriminatory circumstances, with lower wages than white people, were often expelled from pubs and clubs and attacked by young whites where they lived. It was in this context that their culture of origin became increasingly important, not only as leisure, but as a form of affirmation and identity.

It was in 1958, after a conflict between whites and blacks that resulted in the murder of the young carpenter Kelsoe Cochrane by young whites, that the writer and activist Claudia Jones started the carnival celebrations:

> We've been wounded. It's time we began to heal. Why shouldn't we enjoy ourselves. Can't we be proud of what we came out of? It's hard enough to stay alive without punishing yourself trying to forget who we are. Try to see this carnival as a tribute, a way of drawing us together... We'll have our carnival because we deserve it. (*Author's note*: From a speech written for the character Claudia Jones in A Rock in Water, a play by Winsome Pinnock, published in *Black Plays: Two*, edited by Yvonne Brewster (London, Methuen 1989) 87–88. This play was based on detailed research and interviews with those who knew Claudia Jones. For further information see: Marika Sherwood, Claudia Jones (London, Lawrence and Wishart, 1999) 150-162, chapter by Colin Prescod entitled 'Carnival').

The first carnival events were held indoors, "in large halls in London. In the mid 1960s *West Indians*, drawn onto the streets by the sounds of Russ Henderson's steel band, joined in a small community street festival in Notting Dale and the Notting Hill Carnival was born"(Apud Tompsett 2005, p. 46).

If the Notting Hill Carnival began on the streets of London in the mid-1960s, its inspiration, according to scholars, began to develop in the Caribbean in the early nineteenth century, when Africans, no longer enslaved, took to the streets and transformed the *Mardi Gras*, with its own rituals. First, they claimed the right to be on the streets and to occupy this public space, in a way that was previously forbidden. Then, they took to the streets cultural activities that affirmed their distinct identities and heritages "with its ritual drumming and chanting, calinda dance and chantuelle commentaries" (Tompsett 2005, p. 43)

However, although such a carnival has developed in the Caribbean, its significance has changed over time. We agree with Jackson (1988) and Cohen (1982) that such an event with political significance is a contemporary British event with deep roots in the colonial past. To understand its contemporary meaning, Jackson and Cohen affirmed that a knowledge of the history of the Caribbean and the geographical change of British racism is necessary. Like racism itself (Sivanandan 1983), Carnival has changed according to the material circumstances and social relations of blacks in both Britain and the Caribbean. For Miles,

> The meaning of Carnival in Trinidad and in Notting Hill is as different as the meaning of Rastafarianism in Jamaica and in Brixton. Neither Carnival nor Rastafarianism can be understood as a passive cultural import from the Caribbean. Both involve "a creative construction of a new cultural tradition, saturating and modifying culture symbols and practices from [the Caribbean] with a specifically English experience" (Miles 1978, p. 2).

Following Cohen as well, I must argue that the ritual and symbolic aspects of Carnival are not autonomous or independent of its political and economic context, while not reducible to it. The culture is not separable from or opposed to the political; it is fundamentally political.

Carnival in London was born and is still an event of contestation that expresses political and ideological conflicts. According to Henrique da Silva, president of the Paraiso School of Samba,

> The Notting Hill Festival brings great controversy to England because they are not used to carnival. This festival was created by the native Caribbean people who came here one day, created a carnival and the thing grew, because they are habits and culture of a people, and the thing took and lost control. There is no way they could stop because that would be very difficult. The strength of an art is part of a people. It's very difficult for them to stop something that they have no control over. No matter how much they say they're going to do it officially, the people will go there, they'll drink, they'll dance… So the area has become the cradle of their carnival here. There's no way they're gonna go against it, so they're gonna take it and try to organize it.
>
> And they don't have the right recipe to control and organize it. I think they would have to look more at the Brazilian carnival, see what the Brazilian does and how he does it. Because after all, they say that the Brazilian is the king of chaos…it's the biggest chaos, but we fix it. It seems that everything is disorganized, but we understand that[…]
>
> (*Author's note*: Interview with Henrique Dias, president and founder of Paraiso School of Samba -29/04/2017)

The Notting Hill Carnival, today considered the biggest street party in Europe, attracts up to two million people, starting on the night of last Saturday in August with *Panorama*, a performance competition between the national "steel bands". On Sunday morning it opens with *J'ouvert*—a traditional procession where participants cover their bodies with mud to celebrate the beginning of Carnival. Also during Sunday, which is also called *Family Day*, there is the encouragement and participation of children in costumes and who dance to the rhythm of trucks with mobile sound systems or "steel bands". There is also "World Music Stage", with calypso and soca performances promoted by the *Association of British Calypsonians* with visiting and emerging artists (https://www.london.gov.uk/events/2017-08-26/notting-hill-carnival-2017).

On Monday, the day Paraiso presents itself, is the *Grand Finale*, where the audience can have fun listening to different rhythms and sounds that come from steel bands, cars with sound, and the presence of samba schools, which perform with artists and dancers.

If the party has many fans who enjoy these days with great intensity, to the sound of different rhythms, not all Londoners share this feeling. A part of the city repudiates the event. On these days the traffic around Notting Hill is chaotic, the streets, shops and subways close their doors. And the crowd invades and, of course, leaves a trail of dirt on the scene.

The dirt that bothers most of the residents during and soon after the event (*Photo* by Fabiana Lopes da Cunha-2017)

A samba teacher, dancer and member of the *Paraiso School of Samba*, as well as the president of the school, Henrique da Silva, stressed the desire of the population and some authorities in London, to end the carnival or move it: maybe Hyde Park. He also stressed that every year there is less safety at carnival and that this ends up generating more problems: in 2016 some people were stabbed during the event and this was widely reported in the press. I also talked to a young Portuguese man who lived next to the circuit, and she told me that the year before two men had tried to rape her and so she decided, from that moment on, not to stay in the city anymore during those days. The year we took part in the Notting Hill Samba School and Carnival parade, the release of pepper spray on the crowd and Paraiso caused commotion and discomfort among the crowd that crowded the streets where the schools and Caribbean groups passed or rather, stagnated, because it came at a certain moment that no one could leave the place anymore. There were also other incidents in the late afternoon and early evening: three persons were attacked with acid (a type of attack that seems to have become more frequent in London).

But if there is a discussion to end the Notting Hill Carnival, I realized in dialogues with traders in the area, that there is a division in their positioning in front of the event. For some, like the manager of a pub in Bayswater, it is a nuisance because they have to reinforce security, because there are a lot of people on the streets, a lot of drinking and of course, some fights that end up in violent events. I could notice this fear during the party period: many bars and restaurants not only close their doors, but also put wood or other kinds of material protection on the windows, windows and doors. In 2016, according to the Evening Standard, on the last day of Carnival,

316 people were arrested and hospitals welcomed over 600 people. Several people were stabbed, including a 15-year-old boy on Wornington Road. Another 20-year-old boy was attacked with a knife at Porto Belo Road and two other young men at Ladbroke Grove.

> Today's arrests were mainly for drug offences, although there were also 11 assaults on police officers.
>
> A further 60 knives were also seized by police, with another 47 arrests for possession of an offensive weapon.
>
> The London Ambulance Service said its staff had treated 221 people, with 32 requiring hospital treatment. (Evening Standard. 29/08/2016)
>
> (*Author's note:* Available at: https://www.standard.co.uk/news/london/notting-hill-carnival-2016-nearly-300-arrests-as-london-street-party-reaches-finale-a3331846.htmlAccessed on 19/08/2017)

On the other hand, in the same interview there are people who say, the atmosphere was calm. In the same way, the opinions are also contradictory with regard to businessmen and shop owners in the region: if there are those who do not like the event and even close the doors after a certain time, there are restaurant owners, who told me that they take the opportunity to earn a lot of money during these two days of celebration, including charging for the use of the toilets and selling beers at a higher price than usual.

Another point to be stressed is that the associations manage to make their structures viable and maintain them mainly through the help of English institutions, such as the *Arts Council*. In other words, in part, the presentation and spectacle of some groups is only possible because there is financial support from an English institution. In the case of the *Paraiso School of Samba*, the 60 thousand pounds a year are of fundamental importance for them to be able to maintain the rent of the sheds where they keep costumes and floats. Of course, the school has other assets: samba lessons, drums, events and participation in festivals throughout the UK and Europe help to cover the maintenance costs and make it possible for Brazilian artists and musicians to come to the carnival.

The *Arts Council's* support is recent in Paraiso, but the participation of Brazilians with the formation of a samba school at the Notting Hill Carnival dates back to the 1980s, with the *London School of Samba*. From a dissidence of *the London School* would emerge another school, the *Quilombo* (which no longer exists), and in 2002 the *Paraiso School of Samba* was created, also by musicians who were part of the London School. In this group were the Silva family: Henrique da Silva and Mestre Esteves, besides Gisele, Maitê de Oliveira, Damien Mannig and musicians from London and other places in Europe and started to outline the idea of the Samba

School, which started with another musical and cultural project, the *Phoenix*, and organized by some musicians and *sambistas* together with Master Esteves (*Author's note*: Sambistas is the name given to musicians, dancers and people who work or are involved with the musical genre known as samba (*Author's note*: Respectfully refers to a master drummer, the musician who coordinates all the rhythm of the percussion instruments of a Samba School). According to Esteves (Enterview, CUNHA, 2017):

> […] I made many friendships in these wanderings that I made in Europe, each place I made a contact, a friend. That's when the idea of the Phoenix project came to me, this before *Paraiso*, it didn't exist yet. This project would gather the best rhythmists I knew all over Europe, who played drums. Most of them, of course, ended up being from here in London. So I rented a church for us to do the project. The idea was to put a chip, a guitar, a string harmony, a singer. And these guys gathered around 30 or 40 to spend the afternoon playing and singing samba, that was the Phoenix project. That was just one time, just one day. I worked for months, it was no money involved. There were a lot of people involved. We set up a scheme and everything happened, we took beer, cachaça, all the instruments and it was. (…) Actually this project was a way to encourage an improvement in the *London* drum set. It was a way of showing that it could be different. Xavier, who is now a *London* singer, took part in this project and sang with us there. So, actually I did it to show the good things about samba and that it could happen here too. Well, that's where *Paraiso* was born. Why was *Paraiso* born? Actually, *Paraiso* is a dissidence of the *London School*, you know that.

Paraiso over these several years of acting in London has received several awards and its look draws attention because it is always possible to glimpse components of the school with their fantasies stamped on photographs in articles in major newspapers in the English capital during the Notting Hill Carnival period. The carnival association, like all of them, has a symbol printed on its flag, the Phoenix, a reference to its history and its beginning with the musical project we mentioned above. Its colours, red and white, refer to the samba school of Mestre Esteves, *Estácio de Sá*, and also has a pandeiro with the green and pink colours in honor of *Mangueira* (school of President Henrique da Silva). Its symbols are based on the history of part of its founders, the Silva family, coming from Rio de Janeiro and with their links to the samba schools there. As it couldn't fail to have a patron saint, *Paraiso* had Ogun, Saint George, as its patron saint, the same as *Estácio de Sá*, Rio de Janeiro's first samba school and, coincidentally, London's.

During our time in London we did several fieldworks such as visits to the facilities of the *Paraiso School of Samba* (they have two sheds, one to store costumes and the other to make costumes and floats), and observation of the dance and drum classes that take place in two different places: on Tuesdays between 7:30 and 9:00 p.m. at Chalk Farm, Haverstock School, at 24 Haverstock Hill and on Thursdays from 8:00 to 9:30 p.m. at Brixton Recreation Center, Brixton Station Road.

Rehearsals allowed us to get to know the members of the school and how it operates throughout the year, as well as its dynamics in the months leading up to Carnival in Notting Hill. In addition to rehearsals and dance and drum classes, the school participates in events and also promotes some of the school's activities related to Notting Hill Carnival.

One of the main events that the *Paraiso School of Samba* promotes before Notting Hill Carnival is the *Paraiso Carnival 2017 Costume Launch*, which took place this year on May 27 at the Brazilian Embassy in London, with the support of *The Anglo Brazilian Society*. This event takes place every year during this period, with the aim of presenting the carnival costumes so that members and those interested can purchase and parade at the school. We ended up making a video of the event for the school in order to help spread the fantasies and make a historical record of it. (*Author's note*: Available at: https://youtu.be/PqfHgXELP64)

On 26 June 2017, we also went to check out the great party for the choice of the queen and the princess of the school drum set. In this event, at the *Piano Works*, there were several jurors associated with institutions linked to culture and art and also the Notting Hill Carnival.

On 6 August 2017, we went to the *Paraiso Carnival 2017 Enredo Launch*, at the *Piano Works*, in order to observe how the samba-style of the school for the Notting Hill carnival was released.

We also participated in the pre-carnival event, at *Floripa*, where we could feel the thrill of kissing the school flag and having the announcement of the launch of our book by President Henrique da Silva and his brother, Master Esteves. On that day, Henrique asked members of the school to attend our launch event and the seminar to present our research that took place right after Carnival at King's College London.

Among the many moments of participant observation, two of them ended up being more striking and relevant to our work: when we parade from *baiana* (*Author's note*: Traditional dress of the black and half-breed women of Bahia, used in a stylized way in the carnivals of Brazil, which consists of long skirt, lace gown, turban, cloth of the coast, slippers, necklaces, earrings and "balangandãs"—other accessories) and standard-bearer at an event in Primrose Hill and, during the Notting Hill Carnival, when we were able to follow all the preparations and the school parade, following, photographing, filming and interviewing the participants and the event itself.

At the Primrose Hill event, which takes place every year to commemorate the spring entrance to London, we had a really unforgettable and enriching experience. We understood and felt what it was like to parade through the streets of London, the reaction of the public, especially the children in relation to the costumes, the performance and the sound of the school in the neighbourhood. On that day, we also took the opportunity to do interviews before and after the parade, while the members gathered, dressed their costumes (including me) and during the barbecue that was prepared for the community to have a moment of relaxation and also the union of the school. It was a wonderful and very joyful day for the whole samba community of *Paraiso*.

On the Notting Hill Carnival day, we participated in the *Paraiso* concentration, which takes place before the parade. For this, we left home at 7:30 a.m. for the school that *Paraiso* rents to make their concentration and leave for the carnival. Many members who live far away end up sleeping there during the days of the party.

We had already been there in the two previous days, when the drums did their last rehearsals, and on those days, before and during the carnival, we also took the opportunity to interview some important people for our work: Rodolfo Manara, Wantuir, Jerome, Gisele Win (one of the first women in the world to participate in a samba school drum set and one of the founders of the *Paraiso School of Samba*),

Marcelo Oliveira (costume designer of the samba schools in Rio de Janeiro and Paraiso) and Cristiano Morato (also costume designer and partner of Marcelo) and Jefferson de Oliveira, *mestre-sala* of the *Paraiso School* (*Author's note*: Mestre-Sala- a prominent figure in the samba schools, who forms a pair with the flag-bearer).

There we accompanied and registered all the movement and also the last preparations for the event. On the day of the presentation, on Monday, we left together with all the components of the school at 9:00 a.m., and we were able to observe the integration of the leaders and the community of Paraiso with other groups that participate in the carnival, such as *Mahogany* (one of the largest Caribbean groups of the Notting Hill Carnival). As we passed by them it was interesting to observe the reaction they had with the school: they started playing with their instruments from their steel band to *Aquarela do Brasil*. Soon came the response of the drums from the Paraiso School of Samba with the sound of bumbos and drums in the rhythm of Aquarela do Brazil, it was exciting to hear this cultural and musical dialogue between the two groups.

This year, the *London School of Samba* ended up leaving first, followed by another Caribbean group that spread a lot of foam through the streets of Notting Hill, making it difficult for the passers-by of *Paraiso* to juggle in their high-heeled sandals.

The high-heeled sandals amid the foam and invasion of people during the parade (*Photo* by Fabiana Lopes da Cunha-2017)

It is important to point out that, in the Notting Hill Carnival, there is no prior order or script of the groups' presentation and, the great concern of the samba schools is to leave before the *Steel Bands* groups, which with their powerful sound boxes make it impossible for the audience to hear the plot. In addition, most of the way there are no bars (as had already been alerted by interviews with members of the samba school) and, people who are watching end up invading the streets when these groups enter the route, which makes the parade of the samba schools difficult. This is because of the sound of the cars, which is very loud, as we have said, similar to our electric trio groups from Bahia's carnival, and because the revelers go after these trucks to dance to the sound of *soca* and other musical genres from the Caribbean. The samba schools have another proposal, more visual, choreographic, they tell and sing a story through their wings, costumes, floats and *sambas-enredos (Author's note*: It is a samba composed especially for the carnival parade and is used by samba schools in Brazil. In its narrative there is a plot, a story that is told around a theme chosen by the school, and which is different every year) and therefore need this space to be accessible. Throughout the parade the supporters of the school, and I was among them, we asked people to leave so that the floats could pass without hurting anyone. It is a unique experience to see and feel hundreds of thousands of people together having fun to the sound of varying rhythms. It is a mixture of ecstasy and fear. That is because in recent years there has been some pressure from the locals and part of Londoners for the Notting Hill Carnival to end or change places. There have been proposals for it to happen in Hyde Park, as we have said before, but it has not worked. The impression that the samba community has and that has been verbalized in the interviews and informal conversations is that the security and the grid diminish every year, perhaps with the intention that something happens so that the event no longer occurs. In 2017, *Paraiso's* delay in entering the event's circuit ended up disrupting its performance.

We left the concentration around 9 a.m. and the parade started only around 12 p.m. The day was very hot and after 3 h, we already had some components from the school looking for a shade to cool off the heat caused by the costumes, a place to sit to rest the feet of the high heels, also used in the streets of London, and keep the energy to parade in front of the Notting Hill Carnival judging commission. The school managed to parade for only a few blocks and then everything came to a standstill, due to the sound cars and the large audience that follows them, dancing, drinking and eating. When everything stopped, we were waiting for the possible resumption of the parade for hours. As *Paraiso* could not advance with its floats on the Notting Hill Carnival route, around 5 pm, the components dispersed and returned to the concentration headquarters to feed and eat the traditional *feijoada (Author's note*: It is a typical Brazilian dish, prepared with seasoned beans and cooked with salted meats from different parts of pork, sausage, *paio*, charc, bacon, etc. and which, in the Northeast, is added to vegetables), accompanied by beer and *caipirinhas (Author's note*: It is a typical Brazilian drink prepared with slices or pieces of lemon with peel, macerated, mixed and shaken with sugar, ice and cachaça or other brandy (like vodka or rum) and a good *samba*.

Already tired before the parade… (*Photo* by Fabiana Lopes da Cunha-2017)

After hours of samba and waiting for the parade to resume: the tiredness (*Photo* by Fabiana Lopes da Cunha-2017)

The Notting Hill Carnival Route in 2017 (Available athttps://co.pinterest.com/pin/532692080771
82511/ Accessed on 27/07/2017)

4   The Connection, the Samba and Its Translation

Since it began its activities, *Paraiso*, in its history, has always brought artists, designers, carnival artists, mainly from Rio de Janeiro, and has participated in events in the UK and Europe with renowned people in Brazil in the artistic environment and in events accompanying traditional Brazilian samba schools. This has given the school an excellent performance and many awards in the traditional Notting Hill Carnival, created by the Caribinians.

The *Paraiso School of Samba* has a majority of English, Caribbean, French, African and people from other Latin American countries. The number of Brazilians is small. As we have already said, its main source of financial resources comes from an English institution, the *Arts Council*. However, the school also has the contribution of members who participate in dance and samba classes that take place twice a week in different neighbourhoods. These members and samba and carnival lovers also help out and participate in events in London and other cities, also raising some money with their participation.

It is important to point out that many members of the drum set and the school went to and still go to Rio de Janeiro (the biggest link with Rio de Janeiro and the smallest with São Paulo is due to the founders of the samba school being from Rio de Janeiro) during the carnival period in order to learn a little more about samba and carnival. Some of these rhythmists and passers-by even end up participating in parades of samba schools in the special group, the access group, or in the parade at *Intendente de Magalhães* (*Author's note*: Get to know the most popular of Rio de Janeiro's carnivals, made by families in the northern part of Rio de Janeiro-Brazil, which brings together about 200,000 people. A different carnival that serves as the basis for the schools of the Sambódromo access group).

The participation will depend on the ability of the musician or dancer. I interviewed some musicians who have participated or composed in the past, the association (Damien Mannig, Jerome Boumedl and Gisele Win) who were part of the history of the school and they reinforced this need: that foreigners have to go to Rio de Janeiro to learn how to play samba and about the importance of participating in the drums of a samba school in Brazil.

Another important issue to be highlighted are the samba projects (and here I am not only talking about the samba schools, but also about the *rodas de samba* (*Author's note*: The samba circle is a very common manifestation in the cities of Rio de Janeiro, Salvador, Recife and São Paulo, Brazil. The so-called *rodas de samba* do not require great financial detachment and usually gather a large number of people who sing and dance around a table, where the musicians play the instruments and sing) such as *Quintal do Samba* (https://www.instagram.com/quintaldosambal ondon/) *and Realleza* (https://www.instagram.com/realleza.uk/), for example, and a project called *Samba de Bamba-UK* (https://www.instagram.com/sambadebamba _uk/), which aims to educate and entertain, using lives, texts and history to spread content about samba and carnival in Brazil. As for the *Paraiso* projects, they use this musical genre for educational and inclusion work, and this dialogue and bond is stronger, from our point of view, when the school presented, throughout its history, themes and plot that had or have a connection with the history of London or Great Britain. Not by chance, these were the years in which it obtained the most awards and prominence during the event. The school was recognized by LCNH from 2003, however, the years in which it was most awarded had as its theme, subjects related to England or the United Kingdom: in 2005 about Charles Darwin and in 2014 with the History of Tea.

Since its creation, the samba school has developed several projects in primary and secondary schools in London to teach samba and carnival to children. Recently, in dialogue with Brazilian musicians living in the English capital, I learnt that there are still schools offering courses on samba or Brazilian music in their curriculum.

One of the members of the association, Betina, has a very interesting social inclusion project that allows wheelchair users to participate in the Notting Hill Carnival.

The Wheelchair Wing of Paraiso School of Samba-2017 (*Photo* by Fabiana Lopes da Cunha-2017)

There are other connections produced with the presence of Brazilians at the Notting Hill Carnival. Just as other associations and associations did during the festivities, *Paraiso*, in 2017, paid tribute to the families of the victims of the fire that occurred in a building in Notting Hill, the Grenfell, holding a minute of silence before the parade. Many relatives of the victims also took part in the parade of the carnival association and, in front of the judging commission, children who survived the fire released green balloons to greet the souls of those who died in this tragedy.

This connection extends to the language, as the sambas are sung in Portuguese and are made available on the school's website in two languages (Portuguese and English) to facilitate understanding and singing during rehearsals and parades. The costumes are made part in Brazil and part here, in London. The floats are produced entirely in London, but often with reused material from carnivals that took place in Brazil, or in other years, in London. Another interesting point in the 2017 carnival was the plot used by Paraiso: the story of the Caribbean and the Notting Hill Carnival. The storyline was set to the rhythm of samba and sung in Portuguese by Brazilian, English, Caribbean and French musicians and people of other nationalities and places, and celebrated with different costumes and three floats that alluded to the Caribbean and its history with the party. The samba was composed by Dominguinhos da Estácio, was recorded in a studio in Rio de Janeiro and had Wantuir as its main interpreter, who moved from Rio de Janeiro to London with the sole purpose of participating in this event representing the school, which he has been doing for several consecutive years.

It is also important to point out that some British and Brazilians have learned or improved certain skills that they already had through their experience within the samba school, with certain tasks and practices, and use this knowledge, or part of it, as a profession. One of the components of the school, for example, became a designer because of his work with making costumes for *Paraiso*. Others work with sound, event production, as musicians, dancers and/or teachers. There are also the "children of samba"—families that were built from living together at school.

Finally, the connections between Brazil, London, samba and carnival are multiple and have increased over the years.

When we were finishing this text we came across a page on Facebook with the name of *Brazilian Events UK* and in its cover photo, it announced the *Brazilian Carnival* that took place on 23 February 2020. Of the 5,200 people interested, the page provides information that the event took place at The Grand, in Clapham, London, with the participation of at least 496 people. In the invitation of the event, the organizers say that besides sambas that will be played live by experienced musicians or DJs, the audience could enjoy other styles of music such as funk, house, MPB, bossa, forró, axé, pagode, sertanejo, hip hop (*Author's note*: The event announces that the public will be able to hear and dance to the sound of different types of Brazilian rhythms and musical genres and also from other places), among other rhythms and musical genres and, having the possibility of also enjoying the vision of dancers and acrobats, with lots of food and drink with the "flavors of Brazil". It seems to us that Brazilians are seeking to celebrate Carnival, to the sound of different musical genres, on the date it occurs in Brazil, but inside a large nightclub, which includes around 1,250 people, probably because in this period, the climate is very cold in the English capital. We do not know for sure the impact of such an event yet. And whether it will happen again. Mainly because of the pandemic and the problems that large crowds in enclosed spaces can bring to people, to public safety, to London or any other place or capital in the world.

However, we can conclude through the exhibition we have made and using Gilroy's reflection on what he calls "Black Britain" that, as well as the experience of Caribbean migrants to the UK has created and brought "external meanings around blackness, extracted in particular from Black America", Brazilians are making a similar move. The different political identities of the Caribbean, even if they have different histories, with migration, has brought with it this complex cultural exchange and the experience in Europe, particularly in the UK and London, has stimulated the creation of a new identity, the Pan-Caribbean

> became important in the elaboration of a connective culture that attracted these different "national" groups to come together in a new pattern, which was not ethnically marked in the way they had been their Caribbean cultural heritage. Reggae, a supposedly stable and authentic category, provides a useful example here. Since its own hybrid origins in the rhythm and blues were effectively hidden, it no longer meant an exclusively ethnic Jamaican style in the UK, and derived to a different kind of cultural legitimacy from both a new global status and its expression of what might be called pan-Caribbean culture. (Gilroy 2001, pp. 174–175).

I understand that the same movement is being built in London and the UK in relation to Brazilian cultural events, in particular samba and carnival. The connectivity occurs through various types of actions and their dissemination and export is adapted to local needs and to the political dialogue and communities living in the UK, in our case, more specifically, London.

Paraphrasing Mário de Andrade and taking a certain liberty in modifying his phrase: "I am a sambista, playing a *cavaquinho*, accompanied by *atabaque*, *surdo* (*Author's note*: Bass drum), *pandeiro* (*Author's note*: Percussion instrument consisting of a wooden hoop, whether or not covered by a membrane, with openings in the hoop where towels or rattles are placed; drum-basco) and *tambourine* ".

References

Abreu R, Chagas M (2003) Memória e patrimônio: ensaios contemporâneos. DP&A, Rio de Janeiro

Andrade Mário de (1982) Macunaíma. Círculo do Livro, SP

Arantes AA (2004) O Patrimônio Imaterial e a Sustentabilidade de sua Salvaguarda. Resgate 13:11–18

Brzozowski J (2012) Migração internacional e desenvolvimento econômico. Estudos Avançados 26:75

Buñuel L (1982) Meu último Suspiro. Nova Fronteira, Rio de Janeiro

Cohen A (1982) A polyethnic London carnival as a contest cultural performance. Ethnic Racial Stud 5(1):23–41

Cunha FL (2004) Da Marginalidade ao Estrelato: O Samba na Construção da Nacionalidade (1917–1945), ed Annablume, São Paulo

Cunha FL (2008) Caricaturas Carnavalescas: Carnaval e Humor no Rio de Janeiro Através da Ótica das Revistas Ilustradas Fon-Fon! e Careta (1908–1921). Unpublished Doctoral Thesis, Tese de Doutorado, FFLCH/USP. https://www.academia.edu/25816769/Fabiana_Lopes_da_Cunha,_CARICATURAS_CARNAVALESCAS

Cunha FL (2009) Caricaturas Carnavalescas na Careta: Uma Visão do Carnaval Carioca Através da Ótica das Revistas Ilustradas (1908–1918). Geografia e Pesquisa (UNESP at Ourinhos)

2:61–82. http://vampira.ourinhos.unesp.br/openjournalsystem/index.php/geografiaepesquisa/art icle/view/84

Cunha FL (2009) As matrizes do samba carioca e carnaval: algumas reflexões sobre patrimônio imaterial. Patrimônio e Memória (UNESP) 5:1–23. http://pem.assis.unesp.br/index.php/pem/art icle/view/90

Cunha FL (2011) Carnaval X Entrudo: Formas de Regrar o Carnaval no Rio de Janeiro em fins do Século XIX e Início do Século XX. Cadernos de Pesquisa do CDHIS (Online) 24:1–17. http:// www.seer.ufu.br/index.php/cdhis/article/view/13726

Cunha FL (2013) Samba Carioca e Carnaval: sonoridades, identidade, urbes e imaterialidade. In: Garcia TC, Tomas L (eds) Música e Política: Um olhar transdisciplinar, 1st edn. Alameda Editorial, vol 1, São Paulo, pp 201–229

Cunha FL (2015). Sonoridades carnavalescas e identidades: sons, ritmos e diferentes festas no Rio de Janeiro em fins do século XIX e início do século XX. In: Tânia Costa Garcia and José Adriano Fenerick, editors. *Música Popular:* História, Memória e Identidades, 1st edn. Alameda Editorial, vol 1, São Paulo, pp 220–245

Cunha FL (2015) Os -Cordões- entre confettis, serpentinas e lança-perfumes: o carnaval do -Zé Povinho- e as diferentes formas de brincar e tentar regrar o carnaval carioca em fins do século XIX e início do XX - https://doi.org/10.4025/dialogos.v19i2.976. Dialogos (Maringa), 19, pp 565–591. http://ojs.uem.br/ojs/index.php/Dialogos/article/view/33767

Cunha FL (2018) Samba Locations: An Analysis on the Carioca Samba, Identities, and Intangible Heritage (Rio de Janeiro, Brazil). In: Lopes da Cunha, Fabiana, dos Santos, Marcilene, Rabassa, Jorge, editors. Latin American Heritage Interdisciplinary Dialogues on Brazilian and Argentinian Case Studies, 1st edn, vol 1. Springer International Publishing, Berlin, pp 3–20

Cunha FL (2019) Entre Pandeiros, Tamborins, Ritos e Humor: Uma Discussão Entre o Local e o Global e o Patrimônio Intangível. In: Baca Salcedo RF, Tenório Gomes SH, Benincasa V (eds) Arquitetura, Urbanismo e Paisagismo: Novos Contextos e Desafios, 1st edn, vol 5. Cultura Acadêmica, São Paulo, pp 29–61. https://pt.scribd.com/document/441612210/Arquitetura-urb anismo-e-paisagismo-v5-pdf

Cunha FL (2020) Between temple yards and hillsides: Rio de Janeiro's Samba, Its Spaces, Humour And Identity. In: Treece D (ed) Music Scenes and migrations: space and transnationalism in Brazil, Portugal and the Atlantic, 1st edn, vol 1. Anthem Press, London, pp 42–53

Cunha MCP (1920) Ecos da Folia: Uma História Social do Carnaval Carioca entre 1880 e, São Paulo. Cia, Das Letras

Cunha MCP (2002) "Vários Zés, Um Sobrenome: As Muitas Faces do Senhor Pereira no Carnaval Carioca Da Virada Do Século". In: Cunha MCP (ed) Carnavais e Outras F(r)estas: Ensaios de História Social da Cultura. Campinas. Editora da Unicamp, CECULT, São Paulo

Evans Y, Dias GT, Martins A, Souza AS, Tonhati Â (2015) Diversidade de Oportunidades: Brasileir@s no Reino Unido, 2013–2014. GEB (Grupo de Estudos Sobre Brasileiros no Reino Unido) London, United Kingdom

Evans Y, Wills J, Datta K, Herbert J, McIlwaine C, May J, de Araújo JO, França AC, França AP (2007) Brasileiros em Londres: relatório para a campanha De Estrangeiros a Cidadãos, Depart- ment of Geography, Queen Mary, University of London, London. http://www.geog.qmul.ac.uk/ globalcities/reports/docs/brasileiros.pdf

Frangella S (2010) O Made in Brasil em Londres: Migração e os Bens Culturais. In: Travessia- Revista do Migrante, p 66

Gilroy P (2001) O Atlântico Negro, vol 34. Editora, São Paulo, Universidade Cândido Mendes, Centro de Estudos Afro-Asiáticos, Rio de Janeiro

Gruzinski S (2010) O Pensamento Mestiço. Das Letras, São Paulo, Cia

Hall S (2009) Da Diáspora: Identidades e Mediações Culturais. Editora UFMG, Belo Horizonte

Jackson P (1988) Street life: the politics of carnival. Environ Stud 6(2):213–227

Miles R (1978) Between. Two Cultures? The Case of Rastafarianism. Working Paperson Ethnic Relations 10, Research Unit on Ethnic Relations/Social Science Research Council, Bristol

Padilla B (2006) Redes sociales de los brasileros recién llegados a Portugal: ¿solidaridad étnica o empatía étnica?. Alternativas, Cuadernos de Trabajo Social, p 14

Prentice R (2001) Experiential cultural tourism: museums and marketing of the new romanticism of evoked authenticity. Museum management and Curatorship 19(1):5–26

Sevcenko N (2004) Maxixes e Marchinhas: O Ritmo que Invadiu e a Nação que Dançou. In: Cunha FL (ed) Da Marginalidade ao Estrelato: O Samba na Construção da Nacionalidade (1917–1945). Annablume, São Paulo

Sevcenko N (1989) Literatura como missão: tensões sociais e criação cultural na Primeira República, 3a edn. Editorial Brasiliense, Sao Paulo

Sevcenko N (1992) Orfeu Extático na Metrópole: São Paulo, Sociedade e Cultura nos Frementes anos 20. Das Letras, São Paulo, Cia

Sivanadan A (1983) Introdução Desafiando o racismo: estratégias para os anos 80, vol 25 edição: 2, página (s), pp 1–11

Souza GM (1979) O Tupi e o Alaúde: Uma Interpretação de Macunaíma. Duas Cidades, São Paulo

Sheringham O (2013) Transnational religious spaces: faith and the brazilian migration experience. Palgrave Macmillian, Basingstoke

Tompsett AR (2005) "London Is the Place for Me": Performance and Identity in Notting Hill Carnival". Estudos de História do Teatro; Grand Forks. North Dakota 25:43–60

Vertovec S (2007) Superdiversity and its implications. Ethnic Racial Stud 30(6):1024–1054

Fado, Samba, and the Interface Between National Identity, Cultural Heritage, and Tourism on Both Sides of the Atlantic

Felipe Yera Barchi and Fabiana Lopes da Cunha

Abstract In the nineteenth century, fado and samba were developed as musical genres and became popular with the advent of radio and new media in the twentieth century. In addition, under a kind of state patronage, the two styles were elevated to symbols of Brazil and Portugal with the regimes of Getúlio Vargas and Antonio Salazar, respectively. The similarities between the two genres do not stop there, both were protected as cultural heritage and transformed into a tourist resource at the turn of the millennium. This intertwined trajectory holds many similarities on both sides of the Lusophone Atlantic Ocean, and in this work, our focus is on the contemporary processes of patrimonialization and consequent transformation into a touristic resource.

Keywords Fado · Samba · National Identity · Cultural Heritage · Tourism

1 Introduction

Following a comparative perspective, we have analyzed the historical process of the emergence of popular musical genres such as samba in Rio de Janeiro and fado in Lisbon, as well as their conversions into true national symbols of their countries. Fado is a type of Portuguese singing, traditionally associated with pubs and cafés, that is renowned for its expressive and profoundly melancholic character. The singer of fado (literally, "fate") speaks to the often harsh realities of everyday life. Samba is a Brazilian music genre and dance style, with its roots in Africa and recognized around the world as a symbol of Brazil and the Brazilian Carnival.

Having emerged at the end of the nineteenth century in the poorer classes and keeping close links with the suburbs of the cities in question, these styles became

F. Y. Barchi (✉) · F. L. da Cunha
Faculty of Sciences and Letters, Paulista State University "Julio de Mesquita Filho", Assis Campus, Paulista, Brazil
e-mail: felipeyerabarchi@gmail.com

F. L. da Cunha
e-mail: fabiana.cunha@unesp.br

popular at the beginning of the twentieth century and began to reach middle sectors of the two capitals at that time. With the advent of radio, fado and samba became nationwide and substantially expanded their audiences, as well as in the artistic part they suffered changes to adapt to the new mass media. This movement coincided, in both cases, with the rise of authoritarian governments in search of legitimization and popular discipline (Getúlio Vargas in Brazil and António de Oliveira Salazar in Portugal). From the explicit censorship of songs and unpleasant behavior to the regimes, through the sponsorship of artists of convenience, or even through the registration and qualification of "authorized" artists, fado and samba were, ambiguously, hindered in artistic freedom and catapulted to stardom by a kind of state patronage. More than famous, these genres became national symbols.

Once consecrated by the national and world public as the best expressions of being Portuguese and of being Brazilian, fado and samba begin the twenty-first century being recognized by UNESCO as intangible cultural heritage. Behind the evident cultural importance of these styles, a complex web of diverse actors and interests can be seen, among which the potential as a tourist product capable of generating added value and adding uniqueness and originality to Rio de Janeiro and Lisbon stands out.

Maciela et al. pointed out two essential points for the definition of authenticity of a tourist destination: "*truthfulness/legitimacy*, where the site is true, legitimate, unique, exclusive and original; and *history/tradition/culture*, where the site has differentiated or relevant history, tradition and culture" (Maciela et al. 2018, p. 421). The authors also complement that "the perception of authenticity in products and services tends to motivate the interest of choice and purchase" (Maciela et al. 2018, p. 425) and that the two dimensions proposed in the study are guides for scholars and practitioners in the area, considering that "communication elements that reinforce items of originality and tradition add value in the perception of unique and authentic experiences in a tourist destination" (Maciela et al. 2018, p. 425).

We must not lose sight of this dimension of the problem, since "the Organization for Economic Cooperation and Development (OECD) presents tourism as the strategic activity essential for the development of countries" (Maciela et al. 2018, p. 415) and recommends that "assertive management of tourism, in the face of aggressive competition, motivates the development of authentic elements that stimulate the choice of destinations due to unique and genuine experiences" (Maciela et al. 2018, p. 415).

It should be pointed out that the pantheonization processes of these musical genres are not random. The movements that catapult these genres born on the margins to national and international protagonisms have suffered important and decisive interventions from national political powers. There is an intricate, negotiated process that began in the first decades of the twentieth century in which authoritarian governments instrumentalize musicians and music while contributing to the expansion of the audience of these genres, and continues in the twenty-first century with democratic governments aimed at taking advantage of popularly established genres, both for local communities and tourists.

Similar cases can still be seen with flamenco in Seville or tango in Buenos Aires, among other cases (Thimm 2014). Flamenco is an art form based on the various folkloric music traditions of southern Spain in the autonomous community of Andalusia and Murcia while tango is a ballroom dance, musical style and song typical of Argentina and Uruguay, it was born in the impoverished port areas of these countries, in neighborhoods which had predominantly African descendants. For this reason, we have used a comparative approach, aiming to broaden the understanding of the relationship between music, society, identity, and heritage.

As Otilia Lage pointed out:

> Comparative method, comparative history, historical compartivism or comparative method in history are expressions that define the possibility of two or more different social-historical realities, contiguous and/or separated in space and/or time, being systematically compared in order to establish similarities, differences, generalizations and individualizations (Lage 2018, p. 64)

In general, fado and samba were born in the peripheries of port capitals at the end of the nineteenth century and spent the first decades linked to marginal sectors of society. With the advent of radio, focusing both on the aesthetic standard and expanding the audience of these songs, fado and samba gain the middle sectors of their societies, "dignify" themselves and reach the status of national song in their countries, respectively, mainly under the tutelage of authoritarian governments. Of course, this theoretical framework is not intended to eliminate the singularities and specificities of the music.

This option is fundamental, especially to understand the most recent processes of patrimonilization and transformation into tourism assets. In this sense,

> It is interesting to note how the disinterest or inability to construct theoretically an object of study, with the delimitation of its field of analysis and the placement of questions that would allow it to be conducted, has not allowed to discover in similar non-national parallel manifestations that would better help to formulate models or interpretative hypotheses.
>
> The very rare references to the songs of Aristide Bruant or of the French industrial cities, to the Argentine tango, to the Brazilian samba - forms in which certain similarities with fado were felt, did not develop in the search for common elements that shared and eventually became diversified expressions of the same phenomenon (Brito 2003, p. 15 *apud* Bastos 2013, p. 19).

The comparative method aims to give an overall view of certain phenomena that are repeated with some frequency at the same time as it aims to show the local specificities of the same phenomena. In this way, it is possible to contemplate the common trajectory of fado and samba in the midst of the expansion of broadcasting and the action of authoritarian governments—global phenomena of the time—as well as the very particular ways in which artists have taken advantage of state sponsorship or changed their repertoires. In the same way, we can analyze the most recent processes aimed at transforming fado and samba into icons of their cultures in an effort to reaffirm originality, authenticity, and uniqueness.

It is important to point out that these strategies called by placebranding tourists also reveal curious parallels, because in a highly competitive market, it becomes necessary to invest in differentiation in order to create competitive advantage and

The placebranding aims: (1) develop a strong and engaging positioning and image for different audiences; (2) make the site attractive to current and prospective buyers and users of its goods and services; (3) be efficient in providing products and services; (4) enhance the image of the site so that the added value of the site is perceived (Maciela et al. 2018, pp. 415–416).

The aim of tourism managers is to build a strong image to the point of making this locality different from any other so that the simple mention of the name of the possessed city brings out a series of references (Maciela et al. 2018, p. 416). If within this global panorama of the end of the twentieth century and the beginning of the twenty-first, tourist strategies appeal to musical styles to reaffirm the cultural identity of a tourist destination, so as to make it unique in the world, it would be convenient to remember that a similar effort was made at the beginning of the twentieth century by the music industry in general (composers, instrumentalists, record labels, and radios). Menezes Bastos talked about musical events—occurrences—common to the whole Atlantic Ocean region:

> The principle under consideration has also proved fertile in my research on samba, and this has encompassed from the time when the label, such as tango (Sandroni 2001), indicated Latin American dance-music events of African extraction, from Mexico and Cuba to the platinum region (Menezes Bastos 2005b), to the crucial 1930s, when the first and second type varieties of Carioca samba coexist within a field, in the sense that Bourdieu (1976, p. 122) gives this term.

> Finally, I believe that my study of the 1922 passage through Paris of the famous carioca musical ensemble Os Oito Batutas, under the direction of Pixinguinha and counting among its members with Donga, the author of the lyrics for Pelo Telefone (1917) [considered the first samba to be recorded], also brought important evidence in the sense of consolidating the principle now on display: It was in the light city of the 1920s - a world city that was still the "cultural capital of the world" and home to a universe of international (and plurilocal) musical meetings of great dimensions - that the musicality of Os Batutas was able to find development in the direction of an invention of a musical Brazilianness that would only be consecrated in the 1930s. That is, if the Brazilian modinha and lundu were consecrated in Lisbon in the last quarter of the eighteenth century, the maxixe (as much as tango, rumba and jazz) will find its consecration in Paris between 1920 and 1930 (BASTOS 2013, p. 22)

As Nicolau Sevcenko (1983) pointed out, during the *Belle Époque*, a cosmopolitan, liberal, and pacifist current of thought was formulated that understood the Concert of Nations by the metaphor of a diamond, in which each edge represented a nation and each one of them contributed with sui generis products to compose this universal work. In this sense, Menezes Bastos' interpretation is based on the same argument: sharing a common Western musical culture (called world music) each country has strengthened the national elements (musical styles) in order to make them unique. According to him,

> Regarding the *Oito Batutas* and its extemporaneous connection with fado, it is worth considering that the Argentinean press of the 1920's pointed out that this famous Brazilian musical group, in its 1923 tour through Argentina, included what was labelled as fados in its repertoire, labelled by the newspapers as "Portuguese fados" or simply listed with other genres considered "typically Brazilian" (Bastos 2013, p. 22).

On the other hand, it should be noted that even in the 1920s samba was a marginal genre and it would not be too much to assume that groups like the *Oito Batutas* incorporated styles considered more "worthy" or more famous into their repertoire. It is certain that the outlines that would make samba more recognizable would only be clearer in the 1930s and 1940s—a phenomenon that coincided with retail nationalism and was directly linked to the development of the country's broadcasting system and recording industry. A similar chronology is also true for fado in Portugal.

2 Origins of Fado and Samba

It is a consensus among scholars of Fado and Lusitanian culture (Gasparotto 2014; Nery 2004) that he was born in the second half of the nineteenth in the periphery of the Portuguese capital and was initially linked to marginal sectors of society, such as prostitutes and ruffians. A whore, singer, and guitarist who died in 1846, called Maria Severa, is considered the mythical founder of the genre. Around the 1870s fado began to conquer new audiences, but it continued to be frowned upon by high culture exponents such as Eça de Queiroz. The famous author of Cousin Basil (1878) said so:

> Athens produced the sculpture, Rome made the right, Paris invented the revolution, Germany found mysticism. Lisbon that he created? The Fado... Fatum was a God in Olympus; in these neighborhoods it's a comedy. He has a guitar orchestra and cigarette lighting. [...] The final scene is in the hospital and in the enxovia. The background is a shroud (Queiroz apud Gasparotto 2014, p. 85).

In a period of redefinition of the Lusitanian cultural identity, this generation of literate people were more concerned with criticizing the Portuguese backwardness— taking Classical Antiquity as a reference on the one hand and the contemporary French and English Empires on the other—while romantically inspired folklorists sought to invent the "typically Portuguese" rural songbook. In other words, there was no room for a recent genre linked to social degenerates (Gasparotto 2014, p. 85).

It was only in the 1920s and with the emergence of a new generation of fado singers, such as Berta Cardoso, Hermínia Silva, Ercília Costa, and Alberto Costa, added to the state intervention by creating the profession of fado singers (Decree n°. 13.564/1927) and enabling the establishment of fado houses such as Solar da Alegria (1928), that the genre would achieve the status of a national symbol. It is worth mentioning that the above-mentioned decree imposed rules for the appreciation of fado: it recommended silence during artistic performances in order to dignify the music until then associated with troublemakers of all strains and obliged the registration and licensing of the songs, censoring lyrics, but financially rewarding the authors. According to Nery (2004), these measures would plaster the practice of Fado, hindering its main characteristic until that moment, improvisation. Of course, the state tutelage that started in the 1920s and expanded with the Estado Novo would also imply changes. As Pestana (2013) observes, in the 1900s, "a comparative analysis of Fado scores and recordings reveals that while the music industry exercised a moralistic action and control over this critical and ironic dimension, the record industry gave it voice and space for intervention (Pestana 2013, p. 73). Parallel to this process of institutionalization, the development and expansion of recording and broadcasting technologies would also influence the formatting of the genre. As Menezes Bastos pointed out, "phonography captures, constituted and built (and then it is not defined, merely as a vehicle, but as a process) the entire musicality of the planet" (Bastos 2013, p. 24).

However, as evidenced by several Lusitanian emigrants, still in the 1940s, 1950s, and 1960s, fado was unknown in the poorest and most distant rural villages, with a high concentration in the capital. Many of these emigrants reveal to have known fado only in Brazil (Boscarino 2013, p. 179). In this sense, Valente added concerning the emigrants:

> In terms of music, the maxim is worthwhile: among the patricians, everyone thinks they are the same, capable of enjoying fados, of which they know several pieces, capable of singing them whole. Fados from Lisbon, which in other situations would not be being learned and sung… Thus, the Portuguese from all regions tend to assume Fado as "their" music, although it is more characteristic of the Lisbon region. (Valente 2013, p. 169)

The incipient recording industry bet that its product, modern and technological, would find a better market in the capital and therefore directed efforts to Fado, which was already a popular genre on stage, had recognized artists and still good sales of sheet music books (Pestana 2013, pp. 71–72). The radio broadcasting and the record created the conditions for Fado to spread out of the capital and abroad, characterizing it as typically Portuguese in the eyes of other countries since before the Salazar dictatorship. According to Maria do Rosário Pestana, "throughout the twentieth century, the media have made Fado a performative genre icon of Portugueseity, spreading it around the world as a "Portuguese song" (Pestana 2013, p. 74). However, as Heloísa Duarte Valente added, the state action would also be decisive

> In the Portuguese case, there were strategies during the 1950s and 1960s to propagate the regime through the immigrant community, targeting various cultural associations. These would be the privileged channel for spreading the ideology of the Salazarist New State (Valente 2013, p. 167).

As for samba, there are many debates regarding its place of origin—whether Rio or Salvador—and also regarding the classification of sub-genders. However, there is no doubt that samba was born from the poorest and most marginalized groups, mainly Afro-descendants, and although one can still find a background in colonial times, it is also in the second half of the nineteenth century that the term samba became popular and acquired more recognizable features. In the beginning, it was impossible to dissociate samba from its social practice, the "roda"—circle where people dance, sing, and celebrate. "More than characteristic instruments or melodies, harmonies and rhythms, the roda is a factor of greater identification in the early days of samba, where rhythms such as maxixe and others are performed" (Barchi and Cunha 2019). However, despite the relevance of the debates about the origin of samba, our interest lies in the process that leads samba "from marginality to stardom" (Cunha, 2004).

Radio broadcasting and Getúlio Vargas contributed a lot to this. In this way, we justify our option for samba from Rio de Janeiro, since in the federal capital at the time, the genre takes a similar path to fado, that is, it is elevated to the national symbol in the era of radio broadcasting under state sponsorship.

"By Telephone" ("Pelo Telefone", in portuguese), 1916, is the first samba to be registered phonographically, while Ary Barroso's very famous *Aquarela do Brasil* marks the apex of the instrumentalization of samba by the Vargas regime in 1939. These twenty-three years record the "disciplinization" of samba that adapts to the format of records and radio broadcasting, and also loses much of its spontaneity, leaving behind themes such as rascality to the detriment of patriotic exaltation and praise for the work. This is not merely a process directed by the elites, but the fruit of an intricate negotiation, since the old samba community also reaped benefits. As we have already said, it ranges from marginality to stardom.

The trajectory of fado and samba could also be compared to that of argentine tango or Andalusian flamenco, as Tatjana Thim pointed out.

> The appearance of the denominations tango and flamenco for the artistic expressions we know today dates in both cases from approximately the middle of the 19th century. Both tango and flamenco emerged as a musical hybrid in the urban areas of Buenos Aires/Montevideo and Seville/Cádiz/Jerez, marked by marginal social classes, such as workers and prostitutes, or other ethnic groups, as is the case of the gypsies in Spain. (Thimm 2014, p. 22)

In addition to similar trajectories, Bastos suggested a common origin, a shared horizon between fado, samba, and tango:

> In Brazil, therefore, the transformations generated by the binary system lead, on the lundu side, to maxixe and soon to samba, and on the modinha side, to samba-canção and bolero and then to brega music. [...] From this point of view, it seems that what has happened in Brazil in relation to the fado-canção seems to have happened here in relation to the tango (Brazilian), also called tanguinho (Menezes Bastos 1999): consecrated the genre (and its label) as the emblem of another country (respectively, Portugal and Argentina), it no longer has applicability in Brazil. (Bastos 2013, p. 27)

Starting from a shared, international reference, it would become more evident the national particularities, among them the samba.

2.1 Salazar and Getúlio

Born in a small village between Coimbra and Viseu in 1889, António de Oliveira Salazar studied law at the most traditional Portuguese university—Coimbra. There, he graduated in 1914, started teaching Economics in 1916 and got his doctorate in 1918. With a Catholic and conservative profile, Salazar came to power as Finance Minister ten years later in the government of General Óscar Carmona. Soon Salazar would acquire a reputation as a savior of his homeland and be elevated to the post of head of the Council of Ministers, from which he would exercise de facto power for over three decades—without ever having been elected president of Portugal—a position almost decorative in view of his protagonism.

In continuity with the National Dictatorship that erupted in 1926, Salazar institutionalized his power project in the so-called New State from the 1933 Constitution and would remain in power until 1968, when he was absent for health reasons, and died in 1970.

Getúlio Dornelles Vargas was born in São Borja in 1882 and briefly passed through the army before starting his law degree in 1904. His initial political trajectory follows the typical pattern of the First Republic stabilization. He was a state congressman for the first time in 1909 and would be elected several more times, in 1923 he moved to the Federal Chamber and in 1926 he served as Minister of Finance of the São Paulo President Washington Luís. He was elected president of Rio Grande do Sul in 1927 and in 1930 would lead the 1930 revolution that ended the First Republic.

Until 1934, the so-called provisional government was in place and Vargas committed himself to giving the country a new constitution. From 1934 to 1937, he commanded the constitutional government and in 1937 he gave a self-determination by instituting the dictatorship also called Estado Novo which would last until 1945. Vargas was removed from the presidency the following year and was elected senator by São Paulo and Rio Grande Sul and congressman by several states. In the following presidential election (1950), he won the 1951–1956 term, but did not fully fulfill that mandate, committing suicide in 1954.

The growing popularization of samba coincided with Vargas' political rise until 1930, but from then on it was no longer a coincidence, since there was a larger project for national culture and identity, in order to stimulate the adhesion of the masses to Getúlio's government.

2.2   The Evolution of the Heritage Concept

According to Mônica Starling (2011), four models of asset management have existed over the past three centuries: preservation, integrated conservation, rehabilitation, and deliberative governance. The preservationist model, historically the first of them, is marked by the strong action of the state and a certain elitist character that elected products of the erudite culture to be enthroned. Chosen for their aesthetic exceptionality or link to the landmark events of history, the elected objects were taken out of circulation, "frozen". In this model, in force since the consolidation of the National States until the twentieth century, the fundamental concern was the preservation, no matter if the pieces were inaccessible to the general public or were out of their usual context. In the twentieth century other models would appear, the second one, for example, is called by Starling as integrated conservation and aimed at the environment, ambience, and meaning beyond collectible and preservable objects (Sarling 2011, p. 5). In parallel with the changes in the concept of culture, which would now encompass popular manifestations and mass products, in this new conception of heritage "the simple overturning of buildings or urban ensembles gives way to actions aimed at conserving the balance of the urban and natural landscape" (Starling 2011, p. 5).

Originating in Italy in the 1970s—and applied in Spain—the paradigm of integrated conservation

> It served as a theoretical and practical argument for left-wing municipal administrations, and their achievements as a flagship for building a political image of administrative efficiency, social justice and popular participation in urban and regional planning decisions (Starling 2011, p. 6).

In this model, the local communities and the actions of the municipalities were more important compared to the actions of the National States—typical of the previous model. However, if at first integrated conservation aimed at social use, with the advance of the decades there was a loss of social reference and priority was given to the economic results resulting from revitalizations, mainly related to real estate projects. The weakening of this concept would result in the third model, that of rehabilitation:

> In the same way as the conservation model, the rehabilitation model has as its main differential in relation to the previous model, the importance attributed to the development and economic sustainability of the preserved areas. This differential unfolds in new elements and characteristics that become part of the interventions focused on urban development (Starling 2011, p. 8).

The transition of these models is not random, but precisely in a global context of crisis of social democracy and the Soviet collapse, as well as the rise of neoliberalism. For Starling, another substantial difference between the second and third models of asset management would be the constitution of the work teams that comprise these projects: in the second model architects and historians predominated, while in the third the participation of the business community in the decision-making processes is decisive. Indeed, in the rehabilitation model there is an ennoblement of degraded spaces, followed by segmentation and social exclusion of groups that do not fit into the new and high standard of consumption instituted.

Also anchored in the extended heritage concept, the fourth model seeks to include more actors in decision-making processes:

> The most important differential in this model - which we will call deliberative governance - is the inclusion of new actors in the discussion and debate of public policies, which can stimulate greater capacity for negotiation between public and private interests. [...]
>
> Also noteworthy is the integration of professionals from various areas: anthropologists, social scientists, historians, architects, urban planners, tourism professionals, cultural producers and marketing professionals, in order to build a more comprehensive and adequate view of the broad concept of cultural heritage (Starling 2011, p. 14).

The evolution of the notion of heritage and the metamorphosis of the concept of culture coincided in time with the development of the tourism industry—in its global phase—in order to feed back. With the advance of globalization, especially with the integration of markets and post-Cold War trade relations, there is a process of appreciation of local cultures and their typical heritage. In order to face an internationalized culture that advanced barriers with the strength of capital, the stimulation of the valorization of the peculiar products of each culture started to be seen as a factor of commercial differentiation beyond the identity aspects. If in spite of the beautiful beaches Brazil did not have a sufficient hotel network to dispute the primacy of tourist reception, samba could give unique taste to the Brazilian landscape, serving as a differentiation factor. Within this logic are stimulated projects that aim at the "rescue and promotion of culture" which, in turn, are not only submitted to this business logic, but are also symbols of resistance and legitimate and original expressions.

However, it happens that among the different manifestations of popular culture that the different peoples produce, those that had greater tourist appeal ended up receiving more investment from governments. In addition to abandoning "nontourist" cultural practices, this policy raises a serious problem for "elected manifestations": the stoning of samba and fado as tourist products, with exorbitant prices, are now alienating their producing communities (Rodrigues 2016, p. 75; Cunha 2009, p. 53).

Within this panorama that involves the broadening of the notion of culture and the emergence of new models of asset management, as well as the processes of redemocratization of these societies and development of the tourist industry at the turn of the millennium is that we must think about the processes of patrimonialization of fado and samba.

2.3   The Cities, Heritage, and Tourism

The growth of the Fado tourism industry in Lisbon soon creates a problem for the preservation of authenticity. As Rodrigues pointed out:

> Increasingly staged for a foreign public, Fado is trying to maintain its authenticity and giving voice to the award received, through the preservation and enhancement of cultural identity present in the oldest district of the Portuguese capital. In turn, Alfama tries to preserve its traditional and parochial character, as it has always been known, accompanying a new social dynamic due to the presence of outsiders, while trying to avoid its disfigurement (Rodrigues 2016, p. 5).

In the last decades, Portugal has made tourism one of its economic pillars and Fado contributes to attracting tourists. It was recognized as an Intangible Heritage of Humanity by UNESCO in 2011 and has since become one of the country's advertising stars (Souza 2014, p. 192). Traditional Lisbon neighborhoods such as Mouraria, Madragoa, and Bairro Alto, in addition to the consecrated Alfama, observe the proliferation of establishments focused on the practice of fado with an eye on the tourist flow. In this sense, one of the great assets of Fado houses is the way they combine artistic performances with a characteristic menu, maximizing their profits. Despite the financial success, such initiatives generate side effects, as Souza pointed out:

> Fado is part of Portuguese cultural and social life as a whole, and the prices charged by the establishments are often high, chasing away residents and maximizing the presence of tourists who go to traditional neighborhoods in search of the genre (Souza 2014, p. 193)

According to Jorge Mangorrinha, "in tourism what counts is more and more the difference of destinations and products to capture a qualified demand that enhances our tourist offer" (Mangorrinha *apud* Souza 2014, p. 198) and, therefore, in this search for authenticity, both tourists and receiving places, the different interests must be balanced so that tourism does not stifle the culture and it does not lose in authenticity. The Fado:

> It is a free exercise of identification which, although inscribed at the heart of one's emotions [...], can be felt spontaneously and provide a fabulous re-reading of other belongings through ritualized living of "memories coded in action" (Schechner 2006, p. 52 apud Pestana 2013, p. 84).

Therefore, despite the state interventions either in the Salazar dictatorship or with the current tourist industry, fado persists as a popular genre for tourists, Portuguese, or emigrants. It is an undisputed national symbol. As Pestana pointed out:

> Despite the totalitarian projects described above and the efforts to retain the dynamism of Fado, almost one hundred years after the text cited, Fado not only persists in musical life, but also through the voice of fado singers such as Amália, Carlos do Carmo, or Mariza, among many others, as it has extrapolated the national sphere, conquering the space of lusophony, especially with the communities of Portuguese emigrants, and even the worldmusic universe. The current dynamics of Fado is reflected in the recent election to World Heritage. This dissemination and impact on the musical life of the Portuguese was largely due to the role of the media in its dissemination (Pestana 2013, p. 85).

In the year 2000, the heritage of samba and the creation of tourism products in its surroundings gained momentum. On the one hand the initiatives of Instituto do Patrimônio Histórico e Artístico Nacional (IPHAN) in registering three types of samba—the partido alto, the terreiro, and the samba-enredo—and other projects such as the Morro da Providência Open Sky Museum, for example. This museum was idealized by the architect Lu Petersen in order to revitalize the port area with the creation of the City of Samba and the Olympic Village of Gamboa and also with a view to making a touristic itinerary possible. (Cunha 2009, p. 52). To these initiatives, we could add others such as the creation of the Samba Museum, Cartola Cultural Center, and a little further back in time the construction of the Sambadrome of Rio de Janeiro in 1984. These measures highlight both a process of institutionalization and the recognition of popular culture by the public authorities, but they are also actions that foster samba as a tourist asset.

The side effects of these actions are, once again, the increase in the number of shows for the local public, a phenomenon aggravated in Rio de Janeiro by the exchange rate relationship with foreign tourists.

3 Final Considerations

Aguilar Criado presented us with an important conclusion:

> The new dynamics of cultural heritage has as its general explanation the emergence of local values, the strength of the singular, the importance of the different as a noun of the same global logic that leads to more homogeneous cultures. It is in this search for distinctiveness that local culture gains strength by turning its particularity into an added value (Aguilar *apud* Morel 2013, pp. 59–60).

In the cases presented here, we can observe this broader panorama that involves the internationalization of cultural values and practices in the search for the valorization of local cultures and identities in the two time vectors that we address. In the early decades of the twentieth century, we witnessed technology allied with capital and political power transforming local cultural practices into national symbols. Already at the turn of the twentieth century to the twenty-first, we saw the movement to value these roots as an element of identity and economic exploitation.

To see the impact of capital and political power in these two processes does not mean to deny the authenticity of these identity movements or to delegitimize the popular character of these manifestations, it is only a matter of showing the complexity of these cultural phenomena from both a global and local perspective. Cultural agents act, dispute, and interpret phenomena according to their worldview. In the same way that tourist objectification can produce perverse effects in the communities that produce culture, it can mean the possibility of generating jobs within the cultural or tourist sector, offering better working conditions to large contingents that do not enjoy variety in the offer of jobs, and this relationship can still be more visceral.

If as Bastos (2013) pointed out, "phonography captures, constitutes and builds", it would not be unreasonable to think that tourism can play a similar role. In dealing with flamenco and tango, Thimm (2014) explained:

> From the beginning, flamenco dancing and tango were exposed to the tourist trade and, therefore, to the contemplation of many foreigners. The premature tourism, mainly foreign, generated a demand for both artistic expressions and constitutes up to the present time its strongest stability factor (Thimm 2014, p. 40)

Therefore, according to this analysis, tourism can serve as a stabilizing element—strengthening a label, expanding it in space, and conserving it in time—of a musical genre, whether for tango, flamenco, fado or samba, and other genres could still be included such as Parisian Manouche (or gypsy) jazz and Cuban rumba.

There is no automatic strategy to reconcile heritage and cultural interests with tourist and economic ones. It is from this impasse that the possibility and necessity of a deliberative governance model of heritage management emerges, an instrument for safeguarding the heritage that seeks to avoid both the museification of spaces and practices and the elitisation of shows.

The movements to rescue, enhance, pantheonize, and museify the various popular cultural manifestations at the end of the twentieth century took place soon after the redemocratization of Portugal and Brazil, and this is no mere coincidence, since these manifestations also mobilize large human contingents. However, the popular cultural manifestations had to be formatted, filtered, to the point of becoming tourist products under the risk of complete social forgetfulness and abandonment by the public power. The creation of these cultural tourism products coincides with a moment of optimism on the international scene in terms of overcoming the imminent risk of chaos represented by the Cold War, the promotion of peace and multiculturalism by the UN and UNESCO, and the neoliberal wave that changed the classical forms of work and employment and, not least for tourism, deregulated the airline ticket market giving a decisive boost to the tourist industry worldwide.

According to Simoni Luci Pereira:

> The global production of localities has been forcibly rethought, in the sense of what the anthropologist calls translocalities (Appadurai 1997). Many cities may be assuming this designation when they involve global economic processes and flows of cultures and peoples in transit that, if not divorce them from the nation in which they are inserted, problematize the political notions of territory, bringing to light a cultural issue of broader affiliation. [...]
>
> But this process, instead of signifying the annihilation of traditional cultures, emphasizes that the globalized culture is made and re-made from its encounter with localities. According to Ortiz (1994) it is therefore incorrect to speak of a "world-culture" that is supposed to be above national or local cultures, since the process of globalisation permeates all cultural manifestations, which only exists because it is localised and rooted in the everyday practices of individuals (Pereira 2013, pp. 36 and 39).

In this new context, marked by the internationalization of labor in Europe and the exchange of students from developing countries, with circulation of goods and people in figures never before seen, it has stimulated the search for authentic, original, non-massified and non-standard cultural products, noted more clearly from the years 2000 onwards. An extreme effect of this movement would be the emergence of so-called digital nomads and hipster culture. Therefore, in this movement of search for the one, the authentic, public policies of safeguarding cultural heritage emerge, especially the immaterial ones, since the entrepreneurs of the twenty-first century have cleverly understood that successful brands guide their operations by selling consumer experiences and no longer in simple products as before.

References

Bastos RJM (2013) O fado como integrante do sistema de transformações lundu-modinha-fado: Notas para um modelo histórico-antropológico das relações musicais Brasil/Portugal/África in Valente, Heloísa (org), 2013. Trago o fado nos sentidos – cantares de um imaginário atlântico. Ed. Letra e Voz/Fapesp: São Paulo 2013:18–31

Boscarino Jr A (2013) Do Tejo ao Rio de Janeiro: Uma história de fados in VALENTE, Heloísa (org). *Trago o fado nos sentidos – cantares de um imaginário atlântico*. Ed. Letra e Voz/Fapesp: São Paulo 2013:174–191

Barchi F, Cunha FL (2019) Música Popular, Identidade Nacional e Patrimônio Cultural: perspectiva comparada do Tango, Fado e Samba. Anais do II Simpósio Internacional Patrimônios – UNESP: Ourinhos, abril/2019, pp 294–309

Cunha FL (2004) Da marginalidade ao estrelato: o samba na construção da nacionalidade. Annablume, São Paulo

Cunha FL (2009) As matrizes do samba carioca e carnaval: algumas reflexões sobre patrimônio imaterial. Patrimônio e Memória, vol 5, no 2, pp 34–57, Assis, dez

Cunha FL (2013) Samba Carioca e Carnaval: sonoridades, identidade, urbes e imaterialidade. In: Garcia TC, Tomas L (eds) Música e Política: Um olhar transdisciplinar, 1st edn, vol 1. Alameda Editorial, São Paulo, pp 201–229

Cunha FL (2015) Sonoridades carnavalescas e identidades: sons, ritmos e diferentes festas no Rio de Janeiro em fins do século XIX e início do século XX. In: Tânia CG, José AF (eds) Música Popular: História, Memória e Identidades, 1st edn, vol 1. Alameda Editorial, São Paulo, pp 220–245

Cunha FL (2015) Os -Cordões- entre confettis, serpentinas e lança-perfumes: o carnaval do -Zé Povinho- e as diferentes formas de brincar e tentar regrar o carnaval carioca em fins do século XIX e início do XX. https://doi.org/10.4025/dialogos.v19i2.976. Dialogos (Maringa), vol 19, pp 565–591. http://ojs.uem.br/ojs/index.php/Dialogos/article/view/33767

Cunha FL (2018) Samba Locations: An Analysis on the Carioca Samba, Identities, and Intangible Heritage (Rio de Janeiro, Brazil). In: Cunha F, dos Santos F, Jorge R (eds) Latin American heritage interdisciplinary dialogues on Brazilian and Argentinian case studies, 1st edn, vol 1. Springer International Publishing, Berlin, pp 3–20

Cunha FL (2019) Entre Pandeiros, Tamborins, Ritos e Humor: Uma Discussão Entre o Local e o Global e o Patrimônio Intangível. In: Baca Salcedo RF, Tenório Gomes SH, Benincasa V (eds) Arquitetura, Urbanismo e Paisagismo: Novos Contextos e Desafios. 1st edn, vol 5. Cultura Acadêmica, São Paulo, pp 29–61. https://pt.scribd.com/document/441612210/Arquitetura-urb anismo-e-paisagismo-v5-pdf

Cunha FL (2020) Between temple yards and hillsides: Rio de Janeiro's Samba, its spaces, humour and identity. In: Treece D (ed) Music Scenes and migrations: space and transnationalism in Brazil, Portugal and the Atlantic, 1st edn, vol 1. Anthem Press, London, pp 42–53

Gasparotto L (2014) Alma e destino do povo português: O fado como identidade nacional lusa no limiar do Estado Novo (1927–1933). Oficina do Historiador, Porto Alegre, EDIPUCRS, vol 7, no 2, jul./dez. 2014, pp 80–96

Lage O (2018) História comparada e método comparativo historiográfico: problemas e propostas. Atas do Workshop Alto Douro e Pico. Paisagens culturais vinhateiras Património mundial em perspectiva multifocal: experimentação comparada, Porto

Maciela J, Maffezzollib E, Martins E (2018) Autenticidade de Lugar: mensuração e influência na seleção de destino de férias. Revista Turismo em Análise, ECA-USP, vol 29, no 3, set./dez, pp 413–427

Morel H (2013) Buenos aires, la Meca del tango: procesos de activacion, megaeventos culturales, turismo y dilemas en el patrimonio local. Publicar - Año XI N° XV - Diciembre de 2013 - ISSN 0327-6627 - ISSN (en línea) 2250-7671

Pereira SL (2013) Migrações, mundialização, cultura midiática In Valente, Heloísa (org), 2013. Trago o fado nos sentidos – cantares de um imaginário atlântico. Ed. Letra e Voz/Fapesp: São Paulo 2013:32–43

Pestana MR (2013) O fado: Destino e oportunidades do "ser português". Um estudo sobre fado e emigração in Valente, Heloísa (org), 2013. Trago o fado nos sentidos – cantares de um imaginário atlântico. Ed. Letra e Voz/Fapesp: São Paulo 2013:66–87

Rodrigues I (2016) O Fado e a valorização turística dos bairros lisboetas: estudo de caso no bairro de Alfama. Unpublished Master Thesis in Tourism and Education Universidade de Lisboa, 2016, 91 p

Santos R (2011)"Mi Buenos Aires querido": o tango como expressão da nacionalidade argentina nas políticas culturais do regime peronista (1946–1955). Anais do II Seminário Internacional de Políticas Culturais, 2011, Rio de Janeiro, Fundação Casa de Rui Barbosa e Itaú Cultural, vol 1

Souza RN (2014) Um turismo afastado: Uma análise dos usos turísticos e de lazer do fado nos bairros tradicionais de Lisboa. Revista Semina, Passo Fundo-RS, vol 13, no 1, pp 189–199

Sevcenko N (1983) O Cosmopolitismo pacifista da Belle Époque: uma utopia liberal. Revista de História – USP, no 114. São Paulo 1983:85–94

Starling M (2011) Entre a lógica de mercado e a cidadania: os modelos de gestão do patrimônio cultural. Anais do II Seminário Internacional de Políticas Culturais, 2011, Rio de Janeiro, Fundação Casa de Rui Barbosa e Itaú Cultural, vol 1

Thimm T (2014) Modelo de negocio del turismo del baile en Sevilla y Buenos Aires: aspectos de la importancia del patrimonio cultural inmaterial en la gestión de destinos turísticos. Via Tourism Review, no 6, 2014. https://journals.openedition.org/viatourism/706?lang=pt. Accessed 08 March 2020

Valente H (2013) «Para se cantar o fado, tem de saber dar o mordente ! » . Manuel Marques, Adélia Pedrosa e a difusão da música portuguesa na capital paulista in Valente, Heloísa (org) 2013. Trago o fado nos sentidos – cantares de um imaginário atlântico. Ed. Letra e Voz/Fapesp: São Paulo 2013:152–173

Importing the Spirit of Samba. Reflections on Intangible Heritage and the Reproduction of the Cultural Expression of Samba Outside of Brazil

Yago Quiñones Triana

Abstract This chapter analyzes how Samba, as a form of cultural expression, has been particularly developed in Colombia, exploring some facts related to the project of a group of artists who aimed to recreate the Brazilian Samba culture in Colombia, and the challenges they have encountered. This specific expression will be discussed and analyzed through the lenses of some ideas of intangible cultural heritage. Specifically, it has been identified that there is an intention to reproduce something more than a musical rhythm or a dance, but there is a purpose to revive a concept that is difficult to define and that has to do with the life experience related to Samba. This idea of the "spirit" of Samba is what it is meant to be reproduced in Colombia; however, it is important to question if this is possible to achieve, and to identify the cultural barriers that may be found in the cited process. It is also relevant to conclude if the "spirit" of Samba is real or if it is an idealistic construction, and ultimately what is the relation of this case study with the reflections on intangible heritage.

Keywords Samba · Cultural heritage · Carnival · Colombia · Latin america

1 Introduction

Almost two decades ago a group of artists returned from Brazil with the firm intention of replicating in Colombia the experience they had lived with the Samba in Rio de Janeiro. The contact with Samba as a popular culture has been especially remarkable for them, that is, a space where people so much differentiated in terms of social class, age, and skin color could come together and participate collectively in an artistic expression, full of emotion and meaning. This perception made them interpret Samba as a special movement, powerful, capable of messing with deep aspects of the human

Y. Quiñones Triana (✉)
Anthropology, Federal University of Rio de Janeiro (UFRJ), Rio de Janeiro, Brazil
e-mail: yagoqt@gmail.com

Policy, Science, Technology and Society Center-NP/CTS, University of Brasilia (UnB), Rio de Janeiro, Brazil

45
F. Lopes da Cunha and J. Rabassa (eds.), *Festivals and Heritage in Latin America*,
The Latin American Studies Book Series,
https://doi.org/10.1007/978-3-030-67985-9_3

being and having a power with potential of transforming the social environment (SAMBA 2016). This, together with the idea very well spread in Brazil, about Samba being the voice of the inhabitants of the peripheries that finds in the image of the *favela*, the popular neighborhood of the popular classes, the symbol of the place of the excluded, made them create a vision that highlighted these particular aspects. A popular culture then, rooted in the social context in which it would have been born and carrying a claiming and identity message with a strong emotional appeal.

As we can see, these aspects are not directly related to the concrete execution of a rhythm or a dance, or even to any practical dimension of Samba. On the contrary, it is an idea that focuses on a series of expressions that seem to belong more to the immaterial plane and that would therefore be passive to be, in a certain way, implanted in another context, in this case in Colombia. In this elaboration, we see then that it is possible to identify a series of values that are independent of individuals and material factors and focus on a kind of essential repository of Samba. This essence, this feeling, seems to be able to be transmitted and apprehended, regardless of geographical conditions, from a kind of model based on the most striking characteristics of the cultural expression that this group of Colombian artists experienced in Brazil. Speaking in concrete terms: if there is the concrete difficulty of finding, for example, the appropriate musical instruments needed, whether it is possible to improvise in this aspect, whenever the "spirit" remains faithful to what was identified for them.

This idea of the spirit of Samba is difficult to conceptualize or even to verbalize, but it is evident that we are dealing with an ideal image and that it carries a series of values that have a character relatively far from the concreteness of the practice. This spirit of Samba, although without some very specific content, is then the driving force of the cultural project that has been trying to gain strength and space in Colombia so far. Having as paradigm not as long as Samba as such, but a specific interpretation instead of it, it could not be different, since cultural expressions are never fixed models in themselves, but representations of the individuals who interact with them.

The simple idea of seeking to implant a certain cultural expression from one context to another, based mainly on the unique character attributed to such a movement, circumventing evident practical and concrete barriers and relying on a kind of universal force that surpasses borders, is already worthy of admiration. Now, bring this concept into practice in a broad cultural project is even more interesting and, moreover, it raises a series of questions. Among others, it seems valid to ask what kind of interpretation of the Samba phenomenon allows us to identify a grasp of basic values of a cross-cultural character and tendentially opposed to the ephemeral and transitory. In the same way, it is interesting to note how this particular expression of Samba seems to inhabit the universe of the immaterial and defy the threats of mischaracterizing influences, which generally come from interested market pressure or modernization. It is believed that Samba has something more than the simple concrete value of its practice as it appears in the context where it is born, and that it is therefore covered with an intrinsic value that is worth spreading and protecting, because it represents a value that culturally enriches those who come into contact with it. The above hardly allow us to refer to common discourses on heritage, especially intangible cultural heritage. In fact, elaborations on this theme, especially those that differ and criticize institutional and official environments, allow us to problematize an essentialist vision of Samba that permeates the patrimonialist perspective on the subject, a task that we will assume here from the concrete case of the attempt to recreate this cultural expression in Colombia.

2   A Colombian Samba Adventure

The initial facts that gave rise to the phenomenon analyzed here coincide with the particular trip of an enthusiastic group of young Colombian artists who, approximately twenty years ago, decided to take a tour throughout South America, with no fixed budget or itinerary and loaded practically only with their art to show on board an old jeep. This trip, which gains shades of adventure, if we consider the ambition of the project and the precarious logistical and material conditions, it would end up bringing back, many months later, a group of people totally transformed. On the road, couples would dissolve and others would appear, new components were also added to the group, Brazilians by the way, but not all identifiable as human beings, since Samba would also return with them. And it came back in the form of a kind of revelation, or at least initiation, that would not allow the detachment from Brazil to erase a taste that resembled a particular and benign type of virus, from whose contagion no one wanted to be cured and that would give meaning to many of their life choices for years to come.

This interpretation, clearly romantic, but consistent with the feeling of the group at that moment, permeates the meanings of what we could consider a kind of founding myth of the Samba movement in Colombia. It will be the starting point to build the interpretations of what the Samba would be in its relationship with the local context, as well as the models to be inspired in the effort to recreate outside Brazil the cultural expression that they had identified. As it is perhaps already possible to imagine, such an enterprise presents difficulties of various kinds and raises relevant concerns about the real possibility of identifying the immaterial characters of a given cultural expression and "exporting" them, despite the cultural and material peculiarities of the local environment.

After their return to Colombia, the group took various paths and its members began parallel projects. In addition, other artists with relatively independent life stories have also created other artistic and musical proposals related to Samba over the years. Also, projects in other Colombian cities have taken shape with different successes, so it is good to clarify that by talking about Samba in Colombia we are reducing to what has been happening in Bogotá. Even so, it seems attentive and, within the dimensions of the present reflection, to limit our gaze to the relevant projects formed around the figure of Arturo Suescún, better known as "*El Flaco*," a nickname that means "the skinny one" in Spanish. He is an important local artist who participated, of course, in the initial trip mentioned above and who, to this day, he represents a reference of the Samba culture in Colombia. He is a leader who has brought many people closer to this cultural expression and has formed several waves of artists who have given continuity to his teachings. Let us take a brief look at the history of the projects that he has directed and helped to emerge, in order to have concrete elements on which to start the broader reflection introduced above.

Once the group returned, after the South American tour, the first problem, of course, was finding the people who could support this "crazy" idea of trying to do Samba in Colombia. However, the enthusiasm and illusion that the company demonstrates were so great that, very early on, interested artists appeared and ended up infected with the imported Samba virus. Many of them, without ever having stepped foot in Brazil, ended up practically co-opted by the feeling that *"El Flaco"* transmitted to them. In this sense, especially striking for him seems to have been the first straight experience in a Samba school court rehearsal. *"El Flaco"* still tells today with great emotion how was the impact of that first time, the impression of seeing that meeting where rehearsed the community of the Samba School of *Ilha do Governador* (SAMBA 2016). The initial surprise of seeing such a varied group of people together to practice, followed the admiration of realizing the effort they were making, because it was very late at night in a weekday, and none of those workers seemed to complain (SAMBA 2016). Stories like the one before, full of feeling, were present in the initial stages of formation of many people who knew little or nothing about Samba, and who ended up being part of *"El Flaco"*'s" projects.

Thus, a project was born not very well defined, because it was not clear about the forms and methods it had to follow, but that sought to lead to the practice of Samba in Colombia, to replicate the feeling experienced in Brazil. In concrete terms, some musicians and artists were trained in the basic rudiments of the rhythm in question and the necessary instruments were imported or adapted (SAMBA 2016). This initial exercise coincides with the growing demand from the Brazilian community in Colombia and from local sympathizers of Brazilian culture, to enjoy the work of artists who "made" Samba, that is, who worked on this cultural expression in a way at least assimilable to the supposed original source. At this point, the possibility appears and, at the same time, the requirement to create a responsible proposal at the artistic and cultural level, respectful of the basic lines of Samba, but also in a relatively commercial manner. In other words, a proposal that would allow the generation of income and thus represent a form of subsistence for the artists involved (SAMBA 2016). This makes it possible to initially create a group with a particular format that they call as *"batucada,"* similar perhaps to what would currently be a Samba School drums show, that is, a reduced sample of the instruments of a *bateria* (the traditional ensemble of several drums used in any Samba School) and that would accompany the harmony (string instruments) and voice to interpret Carnival sambas. This format, however, did not explicitly imitate the drumming scheme of the show, but simply sought to recreate with the limited material and human resources available, the rhythms and dances of the Samba from Rio de Janeiro.

To this end, an artistic proposal was created that would also be attractive to the lay audience on the subject, for which some visual resources were aesthetically appealed in accordance with the common imagination about Brazil and ended up expanding the repertoire to other rhythms and expressions of Brazil, thus musically diversifying the proposal. With the maturing of the project and the passing of time, the idea of exploring other formats emerges and is created a group that proposes to interpret "*chorinho*" (an instrumental musical genre) and some themes of MPB (acronym to indicate Brazilian popular music and that includes several rhythms). As well as a larger formation, basically of percussion, that would constitute a Carnival "*bloco*," a format that plays Carnival music and parades through the streets. Summed up in a few sentences, this story seems simple and fast, but it is full of the obstacles and difficulties that working with a relatively little known artistic expression in a foreign context can bring. Nevertheless, within the limits of this text, it is relevant to note that, with great effort and through the work of many years, a project has been built that brings together several groups and artists who try the difficult task of making Samba in Colombia.

3   Samba in the Heights of Bogotá

As it can be easily deduced, what these artists cultivate and practice is not "the Samba," but a concrete version of their own interpretation of this expression, strongly adapted to the characteristics of the country where it is done. More specifically, to the city of Bogotá, a mountainous and rather cold region, very far from the coast, with a closed culture, of people generally little expansive and introverted, and in which the population of African descent has a historically irrelevant representation. That is, a cultural landscape that would present itself, in appearance, as incoherent or even opposite to the context that we generally associate with Samba. The result is musical formats unusual to traditional Samba, and practiced in a cultural context that differs markedly from that which historically allowed and encouraged the emergence of this cultural expression (IPHAN; CCC 2014).

More specifically, there is no Carnival or Carnival season in the city and no place or time to the Samba and its expressions to have a privileged stage and thematic climax. We know very well that the poetics of Samba, what it talks or sings about, often has to do with the odd and extraordinary moment of Carnival, but this simply does not happen in Bogotá. We also have the almost total absence of the figure of the "community," that social, cultural, and historical link with the popular sectors, usually through a neighborhood of the city that identifies itself in the practice of Samba and reinforces an ideology of communion and, if not criticism, at least the appreciation of the community as a subordinate agent but one that resists the great flows of power and capital. Even though there are popular neighborhoods in Bogotá, the Samba movement does not target them explicitly, neither as public nor as protagonists. There are some active members who may belong to popular sectors, but there is no "community" subject within the phenomenon as a recognized and evident vector. This fact, on the other hand, is almost logical, since we are talking about a group of artists who work mostly in the formation of other artists to generate a particular cultural expression. There is neither the presence of a pre-existing historical collective on which to rely nor the capacity to encourage it, and it does not seem to be necessary or binding for the purposes outlined by them. In short, some culturally particular spaces are absent, schematically speaking and only to offer two basic examples, such as the Samba school court or "*quadra*" in Portuguese (the place where the members of the Samba School join together) and the backyard of the "*roda*" of Samba (the traditional way of playing Samba making a circle of people), where several elements rich in meaning and symbolism come together and which, in part, justify the recognition of Samba as a cultural phenomenon of national relevance in Brazil, as a cultural treasure worthy of appreciation and value (IPHAN; CCC 2014).

Of course, it seems relatively simple to note these differences between the Colombian project and the model that inspires it, all the more so by making an external evaluation of the movement and based on concepts that are in part alien to the concrete practice of the artists in question. However, it is in this framework of apparent incoherence that the phenomenon discussed here seems to develop, and it does so largely in the light of the divergent specificities. If there is no historically relevant community that gives a popular substratum to the movement, if there are no symbolic spaces for collective encounter and strengthening, and if the artistic proposal should be varied and versatile to respond to the taste of the local public that sustains the project economically, "*El Flaco*" is in charge of introducing an element that, in its vision, guarantees a kind of authenticity of the project. That is, he is constantly concerned in the processes of artistic formation, but also with the example of his own life history, for transmitting an attitude and a conscious positioning about the artistic work, and that brings a message and a vision explicitly popular and historically situated. "To do Samba," in this sense, and beyond the concrete difficulties and incoherencies of context, would mean to try to connect, to enter into the frequency, in the same wavelength of a practice that is only understood from the values and the feeling that it promotes. In a short and simple way, one can say that "*El Flaco*" tries to make Samba a way of life. In other words, it is the immaterial dimension that can be reworked despite the contradictions, there is an invariable substrate that is independent of material and transitory factors that, in his view, can be recreated. In this sense, the project creates links with patrimonial reflection, and specifically with that associated with Samba.

4   Samba and Heritage

Significantly, one of the documents that underpins and promotes the patrimonial declaration of Rio de Janeiro's Samba (IPHAN; CCC 2014), brings a series of testimonies from important figures of this cultural milieu that reaffirm the idea that we are talking about a way of life: "Samba was not important in my life, no, it was my life" and also, "Samba is the essence of my life, because I was born, I always lived in the midst of Samba" (IPHAN; CCC 2014). In the following statement we have the reiteration, with a different term, of the idea of Samba like a virus, something that enters the system and invades it inevitably transforming it: "Samba is part of my life. I was born in samba and in Samba I will die. I carry in my blood the Samba microbe" (IPHAN; CCC 2014). And finally, perhaps already famous, a phrase that sums up the whole subject in a concrete and poetic way: "Samba is a beautiful way of living" (IPHAN; CCC 2014). Here, recognized characters from the world of Samba seem to synthesize the feeling that has been explicitly wanted to redeem in Colombia and that, in turn, justifies its exceptional and priceless character and somehow authorizes us to consider Samba as an intangible cultural heritage.

Specifically, in addition to the grandiloquent and sometimes generic treatment of the official language on Samba as a heritage, which considers it particularly as a symbol of national identity, we can identify a striking feature that appears in institutional justifications and that relates to its community and inclusive character (IPHAN; CCC 2014). That is, the emphasis is given to the idea of Samba offering a space for disinterested fruition and active participation (IPHAN; CCC 2014). In addition to historical considerations, related to the idea that the expressions linked to Samba have reinforced and made visible the popular heritage of African matrix, to the point of being considered even as a form of resistance (Tinhorão 1998), the factor that seems to be most distinguished together with, a generic and universal reading of heritage, is precisely that related to the cultural environment that Samba proposes and provides. Samba is not reduced in this sense to the musical genre, but involves the whole network of social relations and concrete festive instances, such as the Samba "roda," the dance and the parties in family houses (Tinhorão 1998), in which the cultural practice takes shape with specific characteristics. That is to say, it is a congregating effect that is born, of course, from a specific social context but which, precisely because of its inclusive value, is capable of being understood and experienced by outsiders, and fits in with the frequent and famous valuations on the extraordinary meaning of such experience. It is that sense that seems to overcome the most evident barriers and offer a transcendent experience. It is that aspect of improbable definition and difficult explanation that is sometimes verbalized as the "spirit" of Samba, its essence, and it is also that which distinguishes the authentic Samba, the "true" Samba.

In this last sense, we enter fully into the discussion on the patrimonialization of Samba, even if framed in the general context of the official discourse on cultural heritage. It is basically a kind of reification of heritage and it ends up being understood as a "thing" that is found and that must be identified, protected, and transmitted (Smith 2006), a reification that in the case of intangible heritage even brings a paradoxical picture. Specifically, intangible heritage is reduced to things, such as accessories, spaces, musical instruments, among others, which can be used for its expression, or simply redefined as the "intangible values" associated with places and material objects, regardless of the tautology that represents the phrase "intangible values" (Smith 2006). There is a relevant relationship between the "thing" and its value, which allows to speak of good, generically collective, and which acquires such condition from the recognition and assignment of meaning by a social group. There is heritage first of all when there is the valuation of a human collectivity, based on affectivity. This value that determines the resonance of a given heritage is a consequence of its functions of social cohesion, and of articulating the identity of a given group; functions that also determine the adherence of a given community to a given good (Gonçalves 2005, 2007, 2009) (Borges; Campos 2012). All the above would not subsidize with respect to the object (whether material or immaterial), to the thing that materializes such a patrimony, we would have so that before that thing-value, we are talking about value-thing (Borges and Campos 2012).

We see, then, that this is not a matter of materiality, as a matter of fact, in all expressions related to immaterial patrimony, we have to refer to the physical evidence of any manifestation of this kind. When we speak of "thing" when thinking about and defining the heritage, it is not a matter of materiality or not, because it always seems to be decisive, but it is a matter of perspective of concreteness. That is to say, we are talking about the task of fine-tune the perception and the evaluation of what is heritage to challenge contingency and immanence, and so appeal then to a transcendent and trans-subjective dimension.

Authorized heritage discourse assumes that heritage is something that is "found," that its innate value, its essence is something that "speaks" to present and future generations and guarantees the understanding of its "place" in the world (Smith 2006). This perspective then leads to creating the image of heritage as a kind of "encapsulated inheritance" to be safeguarded for future generations, in order to maintain through time its inherent meaning unchanged (Smith 2006). Therefore, according to this official vision, the "essence"—or the supposed meaning inherent in the heritage, the past and the culture it represents—should not be altered or challenged (Smith 2006). That is, it is that content which is independent of contingencies and particular subjects and which constitutes its primary value, in other words, its intended essence. This distinctive and valued sign commonly appears, albeit in a more elaborate manner, in the concepts about patrimony and patrimonialization, which is strongly evidenced when it comes to protection policies enshrined in the safeguard plans. In this sense, it seems innocent to attribute to the concept of heritage the pathologies or triumphs that its use implies; since the mantle of neutrality that surrounds it is a legacy of the humanitarian postulates on which UNESCO acts and the naturalization of its existence (Salge 2014). The cultural heritage or the discourse on heritage is neither innocent nor unitary, and does not explain transcendent realities that were previously present (Salge 2014).

Generally, the identification of a heritage, in a diagnostic tone, is accompanied by a safeguard proposal, which coherently resembles a form of treatment and, like many treatments, is also not free of side effects (Hafstein 2015). In the case of Samba, the situation is no different, together with the granting of patrimonial status, there is a warning about the dangers that hover over the essential aspects that justify patrimonialization (IPHAN; CCC 2014). They fall, in general, in the context of the threats to the heritage proper of the global discussion associated with post-industrial societies and the processes of economic, demographic, and technological transformations (Hafstein 2015). In our specific case, the challenges come from the pressure of economic interests in the cultural field for co-opting the primordial values of Samba in favor of simple commercial spectacularization (IPHAN; CCC 2014). There is then a concrete danger of losing or distorting "something" about what makes the Samba scene exceptional, worthy of value and susceptible to protection and valuation.

5   Colombian Samba and Heritage

What is the relationship between the situation of the Samba within the reflection on the heritage and the specific phenomenon outlined in Colombia? The explicit point of contact of these two contexts is the concrete presence, even if not material, of the "thing." That is, of what makes Samba being Samba, and which is not defined by the places, the instruments, the spaces, and even the people in question. If on one hand there is an institutional demand, supported by an academic effort, to determine the *quid pro quo of* the question of Samba as a heritage; in the Colombian case, there is the commitment to try to transplant, with the means available, such a basic nucleus. In the conceptual area of heritage, the ways and models are sought so that the encapsulated heritage passes on to the new generations without suffering significant alteration, maintaining its meaning; in the Colombian project, the commitment is for not limiting itself to making an external adaptation of a cultural expression, but to offer that "thing" that transcends artistic performance and the generic enjoyment of the public. In both cases we are talking about the conceptually slippery "spirit" of Samba, so ethereal that it looks like a virus, so pervasive that it resembles life itself, so subjective that it is worthless to explain to others.

However, this "spirit" coexists on the historical level with the actual and daily practice in the living spaces of culture. Thus, it has happened that, recently and almost twenty years after the founding trip, some members of the movement in Colombia have had the opportunity to experience closely the reality of doing Samba in Rio de Janeiro, among them "*El Flaco*" itself. In these approaches, it has become evident the preponderant influence of the economic and commercial factor in Carnival and in Rio de Janeiro's Samba schools, and how some expressions have partially lost their spontaneity, moving away from the community ideology, and how economic interests have tried to adapt the artistic content to the demands of the great media and tourism industry. This situation is recognized even explicitly in formal contexts of reflection on the subject of Rio de Janeiro Samba as a heritage. Where it is recognized that the pressures of the entertainment industry have led to a loss of visibility of the traditional matrixes of Samba in Rio de Janeiro, even though they continue to be worshiped by samba players in the communities (IPHAN; CCC 2014).

We have then that the primary reference, the one in theory more faithful and close to the admired ideal, presents cracks in the expected model that would preserve the values assigned to the cultural heritage in question. On the other hand, and in a parallel way, the right experience of Colombian artists on the current Samba scene, especially associated with Carnival, has allowed us to notice some differences in terms of proposal and format, differences from what is done in Colombia in relation to what actually happens in Rio de Janeiro. This is understood, explicitly or implicitly, as what "really is" or "how it is done" or even what "should be done." That is, the format known as the "*batucada*" in Colombia is practically rare in Brazil; the "*bloco*" that exists there has no voice and does not perform Carnival marches and does not have a faithful group of revelers. In short, it is clear that the Colombian proposal is a collection of references that try to adapt to the local context in the best possible way, meeting the preconceived expectations of the public and, at the same time, trying to offer a work that respects the essence of the phenomenon. This may lead one to think that the current greater knowledge of the effective reality of Samba, on a less ideal level, makes the function of the spirit, of the essence collected two decades ago, lose relevance. Perhaps unexpectedly this has not happened, on the contrary, the possibility that some artists may face today the "essence" of Samba in action, only reinforces the teachings and the experience that had been indirectly apprehended through "*El Flaco.*" The right and personal experience simply brings more elements, is more complete and, at the same time, more subjective.

6   The Concrete Essence of Samba Experience

There is clearly a latent conflict between the image of the essence or spirit of Samba and its daily practice and enjoyment associated with historical, social, and economic factors. However, perhaps this is a contrast that also appears in the relationship with Samba in general and associated with the reflection on it in heritage terms. That is, there is a potential incoherence if we consider the unique value of Samba from an ideal vision, which associates it with purity and authenticity, with the impermeability of change and of localized subjects. But if we see its exceptionality in terms of transcendence and the unique richness of experience, we take the concept of spirit to the plane of personal experience, then the inexplicable and indescribable element so often recorded and sung over the years, returns. This spirit that is independent of the quality of the instruments and the configuration of the physical space, which can be transported and recreated wherever it goes, whenever it is possible to make it "happen," whenever it is possible to create the conditions for "that which gives Samba."

It is then necessary to recognize a particular and exceptional dimension of Samba and its practice that promotes spaces and expressions of coexistence that are registered in unique experiences full of meaning for those who are part of them. However, it is risky to interpret such a factor as an essence, for so much prior to our experience and impermeable to it, and that gives reason and defines the true or authentic Samba. With this we end up granting tints of purity to a dynamic expression, and simplifying the historical processes, past and present, that interact with it. This attitude also typifies many approaches to Samba in general, so it is common to find people and even scholars who consider Samba and Carnival, since always and intrinsically, as a characteristic expression of the popular classes and black sectors of society, when we know well that there was a historical negotiation for these spaces and it was not exclusively about dispute and claim (Ferreira 2019). That certain social strata have been protagonists in the consolidation of a cultural expression grants them the right of historical recognition and the privilege of memory, but not the predominance of a kind of unrestricted property. Samba did not appear in an aseptic social bubble and, once it was realized, by its very nature, attracted followers and admirers who, at first seduced, also ended up respecting it and worshiping it sincerely.

Thus, Samba is yes, resistance, but it has not been just that, and when it runs away from this dynamic it does not cease to be what it is, it does not lose its essence. For it is not found exclusively in one place or social class, one specific race or cultural practice. It is useless to look for this essence, because it exists only while Samba is being made, it is not found in a concreteness of any kind, but it is made by human collectives that are, in fact, too fluid. The spirit of Samba exists because it is possible to live it and feel it, and it may seem an ideal when one tries to respect it and define it. But it is not possible to catch it in the defined field of concreteness, because "it is not," but "it is done" in the temporal spaces of its practice. It is possible to feel it, perhaps to tell it, but hardly to define it.

References

Borges LC, Campos MD (2012) Heritage as value, between resonance and adherence. ICOFOM LAM Regional Meeting - Terms and concepts of museology: inclusive museum, interculturality and integral heritage, 21, 2012. MAST, Rio de Janeiro, pp 112–123

Ferreira LF (2019) Opening conference. In: II Simpósio Internacional de Patrimonios: Cultura, Identidades e Turism, 22 a 26 de abril de 2019, UNESP, Ourinhos

Gonçalves JRS (1996) The rhetoric of loss: the discourses of cultural heritage in Brazil. UFRJ Publishing House: IPHAN, Rio de Janeiro

Gonçalves JRS (2007) Anthropology of objects: collections, museums and heritage. IPHAN-MinC, Rio de Janeiro, p 2007

Gonçalves JRS (2009) Heritage as a category of thought: Contemporary essays. In: Abreu R, Chagas M (Orgs.) Memória e Patrimônio, 2nd edn. Lamparina, Rio de Janeiro, pp 25–33

Gonçalves JRS (2005) Resonance, materiality and subjectivity: cultures as heritage. In: Anthropological horizons, Porto Alegre, 11, 23, pp 15–36

Hafstein V (2015) Intangible heritage as diagnosis, safeguarding as treatment. J Folklore Res 52:2–3. Indiana University Press, USA, pp 281–298

IPHAN; CCC (2014) Dossier of Samba Matrices in Rio de Janeiro. IPHAN, Brasília, DF, p 2014

Salge M (2014) El patrimonio cultural inmaterial como principio de autoridad. In: Colombia, Observatorio de Patrimonio Cultural y Arqueológico Opca, Universidad De Los Andes, vol 6, pp 4–7. ISSN: 2256-3199

SAMBA in the heights (film) (2016) Direction: Yago Quiñones Triana. Cuatro Perros Producciones, Bogotá (47 min)

Smith L (2011) The "heritage mirror". Narcissistic illusion or multiple reflections? Antipode. J Anthropol Archaeol 12:39–63. ISSN:1900-5407

Tinhorão JR (1998) Social history of Brazilian popular music, vol 34. Editora, São Paulo

https://www.youtube.com/watch?v=TRvPHMc8A1M

Schools of Samba in the South Brazilian Border: Circuits and Translocal Exchanges in Carnival Cultures

Ulisses Corrêa Duarte

Abstract Based on a multi-situated ethnography in the Pampas region, developed on the triple border in southern Brazil between Argentina and Uruguay, we will reflect on the carnivals of the schools of samba that occur beyond the center of the country. We will analyze the social-cultural importance of the carnival in Uruguaiana (Brazil) and the consequent impacts of the festival with local politics, the tourist sector, and the economic dimension of the event in the region. The extended carnival calendar of schools of samba in the Pampas promotes a ritual time of the festival that allows for extensive exchanges and translocal negotiations between the carnivals on the border and the carnivals of schools of samba in Rio de Janeiro. The carnival circuits in the Pampas allow us to think about hybridism, globalization, and the cultural identities produced by popular cultures on the margins and beyond the borders.

Keywords Schools of samba · Carnival · Region of pampas · Border · Globalization · Brazil

1 Introduction

The Brazilian carnival has consolidated itself as a party of great relevance in the national calendar, and it has become one of the greatest socio-cultural references abroad. The artistic form of the schools of samba based on the Brazilian carnival is the most celebrated, popular, and requested for the representation of the country in international events, besides being considered the best developed in its artistic potential since the elevation of the samba of musical genre to national symbol in the first half of the twentieth century (Cunha 2006).

State policies fostered the consolidation of the Brazilian identity in the 1930s and led to the country's dissemination of carnival groups that gathered around the associative form of schools of samba, a format that underwent modifications overtime

U. C. Duarte (✉)
Federal Fluminense University (UFF), Rio de Janeiro, Brazil
e-mail: ulissescorreaduarte@gmail.com

© The Author(s), under exclusive license to Springer Nature Switzerland AG 2021
F. Lopes da Cunha and J. Rabassa (eds.), *Festivals and Heritage in Latin America*,
The Latin American Studies Book Series,
https://doi.org/10.1007/978-3-030-67985-9_4

after its historical genesis in the former capital of the Brazilian Republic: Rio de Janeiro.

In the processes of dissemination to other social and geographic contexts, the Brazilian samba and the carnival of the schools of samba, in its version of an improved spectacle, took advantage of its inclusive potential and the cultural transformation of its organization, which adapts to each new place where it was installed and expands to new regions and beyond political and national borders (Duarte 2016). The potential for dialogue in the schools of samba's productions and for promoting a series of intercultural relations between the groups involved in the artistic production of these associations results in their ability to hybridize and hold an event in multiple locations weaving a network of negotiations and meanings (Comaroff 2003).

I developed an ethnography over seven years in the carnival of the triple border between three countries, the Pampas[1] gauchos: located between Brazil, Argentina, and Uruguay. In the region of the Pampas, the carnival is held over the three months of summer, and its main characteristic is the great flow of materials, professionals, and ideas about the improvement of the carnival that circulates among the cities that hold parades, in addition to the strong dimension of globalization of carnival culture as we will analyze throughout this text.

2   Carnival in Uruguaiana on the Pampas' Triple Border

Uruguaiana is located in the extreme west of Rio Grande do Sul's State, about seven hundred kilometers from the State capital, Porto Alegre. It is the largest city in the region of the western border of Rio Grande do Sul, delimited by the Uruguay River that separates it from the Argentine city of Paso de Los Libres through an international bridge opened in 1945. The city has in its territory the sixth-largest cattle herd and the largest planted rice area in the State, which characterizes its economic focused on agribusiness.[2]

The city has one hundred and twenty-five thousand inhabitants, according to the 2010 Brazilian demographic census. Uruguaiana's carnival was considered the largest and most important carnival in the Pampas, considering the number of audience and visitors. Besides this fact, Uruguaiana has the most traditional carnival associations in the region, with large number of participants, outstanding investment in the purchase of carnival materials, and hiring samba players from other carnivals as attractions of the event.

---

[1]The Pampa is a natural biome with undergrowth vegetation, small shrubs, and relief that has hills that are not very rugged and suitable for agricultural activities, such as cattle and rice planting.

[2]According to the 2017 Agricultural Census of the Brazilian Institute of Geography and Statistics (IBGE), there are about 323,000 head of cattle in the municipality of Uruguaiana. According to the 2011 "Temporary Crop" census of the IBGE are about 734 thousand tons of rice (in husk) harvested in the municipality.

The positive result in the numbers of local commerce, hotels, and opportunities in the tourist sector in the 2018 carnival was released by the Municipal Secretariat of Economic Development[3]:

> The statistical report of Carnival, organized by the Tourism Board of SEMUDE, made through research with companies and specialized sectors, showed that the hotels chain of the city (1,630 beds) reached an occupation of 95%, making a total of 1,548 beds used by tourists. In addition, another 55 beds available were completely used. In the hostelling sector 35 beds were available and 31 were occupied, reaching a 90% occupancy rate.

> In the dynamics related to transports, the Uruguaiana Bus Station noted a 10% increase in the flow of movement during the period. Together with the company that has the largest number of daily operations at the Bus Station, the flow of tourist movements recorded a 30% increase compared to other periods.

> At Rubem Berta Airport during Carnival Week, all flights had full capacity to Uruguaiana.

> In all 200 street vendors were registered to act in the sale of food and beverages around Avenida Presidente Vargas. The sales of the first night of Carnival reached the average value of R$350.00 per seller (U$100,00—equivalent value in American dollars at that time). On the second night (Friday) this value rose to R$500,00 per seller (U$142,00). On the last night (Saturday) the average sale value fell to R$300.00 per seller (U$85,00).

For the Uruguaiana's people, carnival was considered the most important festival of the year. The great expectation generated in the tourist and commercial sector in the city, with the arrival of visitors and hired for the carnival, as well as the involvement of all social classes and neighborhoods of the city, transformed the urban environment into a carnival great and popular atmosphere. The schools of samba carnival extended to other municipalities in the Pampas along the Brazilian border and marked a period of massive participation in the carnival associations that were facing a ritual dispute (Turner 1974), established in a competitive cultural practice incorporated as a locally established tradition.

In the Pampas region, considering their border cities, the carnival parades extended throughout the summer months on the calendar. The only one of the three municipalities that celebrated the party on the weekend of their official international holiday was the Artigas carnival. In recent years, the Paso de Los Libres carnival celebrated its festivities on the four Saturdays before the carnival holiday, over a month of parades. In Uruguaiana, the carnival was always scheduled three weekends after the carnival holiday, during the Lenten season.[4]

The extended carnival calendar between the months of January and March in the region results in the non-coincidence of the dates of the festivals, and allows the circulation of samba players and participants, encouraging the exchange of objects and materials among local carnivals and opening the possibility of hiring samba players from the other parts of the country, especially Rio de Janeiro's carnival, a fact that has intensified when the carnival established out of holiday calendar.

---

[3] Available in em http://www.uruguaiana.rs.gov.br/pmu_novo/veiw_noticias/2814 in February 2019.

[4] Lent is the Christian period of righteousness and penitence during the forty days before Easter. It begins on Ash Wednesday, one day after the end of carnival. It is important to remember that carnival is a Christian holiday that was adopted in colonial Brazil from European religion influences.

In 2005, the school of samba Os Rouxinóis, the city's five-times carnival champion, was prevented from using its headquarter because of a lawsuit filed by neighbors to its space. The complaint was filed in the Public Prosecutor's Office and brought the complaint of an alleged transgression of the noise limits, in specific legislation, during party nights. The lawsuit promoted wide debate in the city, and there was the threat of cancelation of the parades by the League of Schools of Samba (LIESU). Sanchotene Felice, the mayor elected for a four-year term (2005–2008), led the agreement between the schools of samba and the Public Ministry at the beginning of his first term. A new municipal law was enacted allowing for pre-carnival rehearsals fixing the noise limits for carnival associations.

As a solution to the 2005 carnival case, the City Hall agreed with the League on a new date for the festival, three weeks after the scheduled weekend. The change of date generated strong repercussion, many criticisms, and doubts in the participants. However, in its first year of accomplishment, Uruguaiana promoted a great attraction of samba players from cities where the Carnival had already been finished.[5]

Since then the format, previously improvised, had a great success and has become definitive. In addition to hiring highlights of Rio de Janeiro's carnival, the possibility of purchasing carnival materials and objects from other carnival poles was observed for the reuse and qualification of local parades.

With the success of the carnival out of season, the same strategy of carnival production was inaugurated in neighboring cities in the Pampas region, in an attempt to participate in this carnival circuit, making use of the carnival labor hired from the center of the country, which was initially attracted to Uruguaiana. Besides the possibility of receiving sambistas and visitors from other cities, after the finishing of the carnivals of their respective cities. The possibility of adjusting a carnival calendar during the summer months in the region helped to form a solid carnival circuit in the Pampas region, increasing the exchanges of people and carnival materials that moved among the municipalities in this extended calendar.

In the first three editions out of season, the League of Schools of Samba of Uruguaiana (Liesu) organized the parades of the associations and the competition, while the City Hall was responsible for building the physical structure of the sambodrome, bleachers, and boxes constructed at each edition. The 2007 carnival was the first edition of the competitive parades in which twenty-two jurors from Rio de Janeiro's carnival judged the carnival. Liesu understood that the formation of a jury completely outside the city, composed of professionals and people linked to the largest schools of samba carnival in the country, would bring greater credibility to the competition in contrast to the contested results in previous years. In addition to this aspect, the Rio de Janeiro's jurors brought the technical demand and standardization of quality criteria in the parades according to their carnival parameter that could qualify the level of competition in the following years, according to the League.

---

[5]In the first edition of the carnival out of season in Uruguaiana, some samba players of Rio de Janeiro's carnival were hired by the Os Rouxinóis, which caused a great impact and repercussion on social medias.

Between 2008 and 2009, an unsolved political conflict between the League and the City Hall produced a serious crisis among the main organization of the schools of samba parades. The point of greatest escalation was the rules for the division of the spaces to be sold on the sambodrome. Newton Gomes, the president of LIESU at the time, reported that most of the revenues of the association came from the sale of boxes and bleachers tickets on the parade days.

With the growth of carnival in that period, LIESU abandoned the oldest amateur model to construct the sambodrome, always improvised and with rather precarious bleachers. When the City Hall became the event's main sponsor, and responsible for the organization, the administration of all services and physical structures of the sambodrome was professionalized.

The interest in taking over the cambodrome's construct increased the City Hall's participation in the management of carnival throughout Mayor Felice's term. The total number of boxes built in the City Hall's project was prorated among parts: ten percent for the League to pay for its physical organization structure (the League's office was located in private offices of its presidents); a part for the City Hall that transferred on to some educational institutions, chosen social projects, and a privileged space for public authorities (special boxes). Most of the boxes and bleachers' tickets were destined to the schools of samba, which resold the spaces and collected their main income for the production of the parades.

With the end of carnival 2009, the crisis between the two entities became unsustainable. Jair Rodrigues, the new president of LIESU and Mayor Sanchotene Felice exchanged public accusations in radio programs and newspapers. The League considered that the City Hall intervened in the artistic direction of the show, which should be its exclusive responsibility. A change in the approved project of the event was made without its consent. Eighty City Hall boxes were installed in the sambodrome, dozens of them not foreseen in the project. Many of these boxes were donated by the City Hall to its employees, students from public schools and charities, without the agreement of the League, which attributed to the City Hall a strong loss in the commercialization of its quota of tickets due to the courtesies granted.

A Parliamentary Inquiry Committee was established soon after carnival by the City Council, with many denunciations from the president of LIESU against the City Hall to verify some irregularities in the use of carnival resources (the carnival trial). The League questioned the municipal administration in the case of donated tickets and boxes added to the original project. The Uruguaiana City Hall charged the league to account for the public subsidies provided in the ticket sales, which was in delay, further increasing the crisis and accusations between the two entities.

The carnival 2009 ended up being the last carnival in which LIESU and the City Hall shared responsibility for organizing in Uruguaiana. Mayor Felice, justifying the better distribution of public funds, regulated the sale of tickets delegating the operation on to the Uruguayanian Trade and Industrial Association (ACIU) since the sale of tickets was prohibited to public agencies. Aciu's board of directors was composed of Felice's political allies, such as the vice-mayor. The schools of samba had an immediate increase in revenue with the control of commercialization and apportionment by the City Hall, besides no longer needing to be involved in the work of ticket sales.

The City Hall's hegemony in command of the festival made it possible to control the provisional sambodrome's project, the marketing of tickets, and the distribution of fees to the associations. Ticket revenue subsidies grew substantially in the following years since a new policy of price and redefinition of special areas implemented by the organizers with the sectorization of the avenue spaces.

The City Hall took upon itself the task not only of the construct of sambodrome's infrastructure, but also became involved in the artistic organization of the parades with the appointment in August 2009 of a Carnival Commission linked to the Mayor's Office.

The carnival trial ended up resulting in a final report presented in July of the same year; the commission understood that there was no mismanagement in the organization of carnival, and the investigation was shelved by three votes in favor (from the government councilors) to two opposing (opposition councilmen) among the members of the investigation. The crisis was suspended. The City Hall has temporarily won the battle to control the carnival, moving the League away from holding the event after carnival 2009.

Thus, Uruguaiana City Hall began to form its own commission annually to deal specifically with the carnival organization: the artistic direction of the festival, the rules of the parades, the carnival jury, the infrastructure of the sambodrome, the security, the food and health services, the sponsorships, and the commercialization of tickets for the nights.

The festival management by the City Hall, with the structure set up by a carnival commission, took place until the year 2016. The political and economic crisis that affected the country, and the consequent fiscal deficits in the municipal treasury, threatened the holding of the 2017 carnival edition. LIESU was unable to rearrange itself. A new Association of Schools of samba was created for its political reorganization. The solution found was the signing of an agreement between the City Hall and a producer company from Pelotas, Bah Entertainment, which modified the ways of funding the festival. The City Hall started not to invest direct resources in the carnival, besides its mobilization in basic public services during the days of parades (security, transportation, and traffic) and its important institutional support.

The hired company was responsible for the infrastructure of the sambodrome and the commercialization of the spaces and the arts organization with a new entity created: the Association of Schools of Samba of Uruguaiana (Asesgru). The same carnival management formula was repeated in 2018, this time with a readjustment of the fees per school of samba, one hundred thousand reais by year (almost U$30,000 at that time).

In 2019, a local supermarket chain called Baklizi overcame Bah Entertainment and set up an artistic production company to promote the carnival in the same way as in previous editions: without the direct participation of the public authorities in setting up the infrastructure for the parades and responsible for the grant to the schools of samba. In 2020, each school of samba received a quota of one hundred and one hundred forty thousand reais to produce its carnival (U$40,000). The management model of the festival has been well established and will probably be repeated in the coming years.

3 Carnival in Artigas (Uruguay) and Paso de Los Libres (Argentina)

Paso de Los Libres is located on the opposite bank of the Uruguay River from Uruguaiana's view. Both cities are connected by an international bridge, one and a half kilometers long, the main route connecting the two countries. For this reason, Uruguaiana has many companies to storage and transport cargo, considered one of the busiest parks in Latin America. The population of Paso de los Libres, about fifty thousand inhabitants, considers carnival its main popular festival. The city's carnivals proudly stated that Paso de los Libres was "la cuna del carnaval argentino", it means the birthplace of the country's carnival. The history of Paso de los Libres' carnival, from the ballroom clubs to the street carnival with the predominance of associations of samba, a Brazilian musical genre, was due to its geographic and cultural approach to the neighboring country, reinforced after the construction of the bridge.

The city of Artigas is located one hundred and twenty-five kilometers from Uruguaiana. Capital of the Uruguayan Department of the same name, Artigas, the city has about forty-five thousand inhabitants and is located next to the Brazilian city of Quaraí in the north, separated by the river of the same name. Many Brazilian visitors are attracted to Artigas by its electronic, cosmetics, and perfumery commerce through its free tax stores.

The carnival production process linked to the schools of samba in the Pampas region was intimately associated with the existence of this extended ritual time and the traffic of people and carnival objects that can be traded within the region. The carnival of the schools of samba is a popular festival with a large number of participants involved in the production of the parades on the triple border. We have already seen that the carnival calendar of Uruguaiana, Paso de Los Libres, and Artigas are not coincident. The carnivals of the Pampas crossed paths cultural and economic exchange circuits through people and things, intensifying the possibility of meetings and negotiations for the qualification of their cultural product.

With small variations, the structure of the parades, the musical genre, and the forms of presentation are similar on Brazil's southern border. As the years went by, Paso de los Libres and Artigas' associations got closer and closer to the art forms of the Brazilian schools of samba, and with them, they constantly dialogue in search of their improvement. The art forms of a school of samba are characterized by a standardized series of elements and minimum requirements that each association needs to present in order to adapt to the rules of the parade and to assert itself as a carnival association within the parameters of samba.

Thinking about the characteristics of the schools of samba, we can identify in Paso de Los Libres and Artigas that the parades happened entirely to the sound of a single samba-plot sung from the beginning to the end of each performance, being composed to be told as a narrative previously established and renewed each year. Many times the samba-enredo was made by composers of the Brazilian carnival. Likewise, the use of brass instruments was forbidden by regulation in schools of samba in Brazil for decades. This fact is not fully followed by other Argentine and Uruguayan cities that have carnival parades with samba influence with some other types of instruments on their musical group playing different songs and rhythms during the parade.

An important characteristic of Artigas' carnival was that the sambas presented in the parades were presented in Portuguese. Usually, these songs were signed by Brazilian composers hired for such demand in many carnival groups. With an intense commercial relationship with Brazil, and with the great insertion of Brazilian media products in the city (such as television, radio, and music), the assimilation of the Portuguese language was a consolidated phenomenon in the Uruguayan city. In basic education, students learned Portuguese and the language of the neighboring country was a specific subject in the elementary school years.

In Paso de los Libres, schools of samba performance with a front commission, the Baianas group, the old guard group, and a couple of master room and flag bearer. As distinct elements of the Brazilian carnival, characteristic of the Argentine carnival, there were the "bastonera" (a highlight that danced with a stick, an influence of the local military bands), and the queen of the institution (a high luxury highlight that paraded in a float exclusive to the character). The plots in Paso de los Libres started to develop a completely narrative a few years ago—with beginning, development, and finalization—in a way adapted to the Brazilian carnival. It is important to note that schools of samba organization, through the storyline—linking a narrative in facts to be told (with allegories, components, and costume groups)—is the narrative structure proper to Brazilian groups in a competition.

The Artigas carnival was widely adapted to the Rio de Janeiro's parameter. The schools of samba presented all the requisites and elements that existed in the regulations of the neighboring country. The narrative storyline, front commission, baianas group, dancers group, drummers, floats, old guard group, and samba-enredo from beginning to the end of the parade nonstop. As the carnival of Artigas coincided with the dates of Rio de Janeiro's carnival calendar, the participation of samba players from Rio was restricted to the pre-carnival period. Therefore, the Artigas carnival groups made many hires of regional samba players and carnival artists, mainly from Uruguaiana and Porto Alegre, the largest carnivals in the state.

The group of jurors of the carnival of Artigas and Paso de los Libres was also hired from Brazil, from other carnivals in the state of Rio Grande do Sul. The evaluation made by Brazilian jurors in the carnivals of Paso de los Libres and Artigas made the forms of interpretation of the samba and dance issues and the appreciation of the visual terms approach the aesthetic view and standards of evaluation proposed to Brazilian samba groups. The Brazilian jurors judged the carnivals of neighboring countries based on their understanding of the carnival culture of their contexts of origin, which made the Artigas and Paso de los Libres schools of samba adjust to the criteria adopted in Brazil, gradually reducing the differences and particularities of their own art forms.

4 The Pampas Carnivals and Their Producers: Rita Maidana and Jéfferson Lima

Some carnival professionals from Uruguaiana circulated among the carnivals of the region and consolidated the social networks, exchange of goods between carnivals, and opening new opportunities to develop their activities. This is the case of Rita Maidana, carnival artist with several works in Pampas region, and Jéfferson Lima, from Nova Friburgo (Rio de Janeiro State). Jéfferson was engaged in the Uruguaiana schools of samba since the samba-enredo music composition to all plastic production of a carnival parade.

Rita Maidana is a carnival artist from Uruguaiana who works in other cities of the region such as Itaqui (Brazil), Bella Unión, and Artigas (Uruguay). Rita was one of the promoters of the consolidation of the carnival flows between Rio de Janeiro and the Pampas region, in the first year that the carnival became out of season in Uruguaiana in 2005. During the 1990s, she was responsible for the plastic production of the school of samba Unidos da Ilha do Marduque, besides being the general director of the costume groups, researcher and plots writer, and worked to construct the floats and the costume workshops.

In a meeting of Ilha do Marduque assembly in 2004, Rita tried to present changes in the conception of carnival and in the work techniques in her school of samba. She intended to create a specific post of coordination in the various plastic-visual sectors of Marduque: the function of carnival arts director. The arts director, in Rita's conception, would be the key post for the organization and idealization of the whole school of samba parade.

The definitive notions about the arts director represented by a specific professional, the coordinator of all plastic and narrative sectors (writing the plot, drawing the elements, construct the floats and coordinate the costumes workshops), did not exist in Uruguaiana at that time. Before, each creative sector was led by one person, without a general director of all areas.

In the preparations for the carnival 2005, Rita was sent by Marduque school of samba to the most important carnival in the country in search of professionals to seek resources to complete the plastic-visual project. In Rio de Janeiro, Rita, in possession of the drawings planned for the costumes, looked for a sewing workshop to produce the costume groups. The use of sewing techniques, plastic solutions, and materials unknown in the southern border of the country, made the hiring of a workshop specialized in carnival promoted a great advance in the visual quality of the parades in Uruguaiana. The increase in the quality of costumes and materials would become essential in the following years, given the qualification of the plastic projects of costumes made with new solutions acquired.

Another task for Rita on her trip to Rio de Janeiro was to look for options for the creation of Marduque's samba-plot that year. Rita got in touch with composers from Rio de Janeiro carnival. A song writer who had been champion at Beija Flor de Nilópolis school of samba that year, known as Wanderlei Novidade, was invited by Rita to compose the music for the Marduque's parade. A few days later, Rita received a copy of a compact disk with lyrics and samba melody that she considered of good quality. When she presented it in Uruguaiana, the school of samba approved the music of the Rio composer without competitive contest as it was common, and invited him to the carnival, with all expenses paid and a benefit to be paid. Besides Wanderlei, a group of musicians and performers from Rio de Janeiro carnival was hired to work. Marduque, as well as many Uruguaiana groups, in those early years of the out-of-season carnival, began to work frequently with composers, singers, dancers, and professionals from the Rio de Janeiro.

The hiring of professionals connected to the Rio carnival for key functions in the parade took place in Uruguaiana in the following years, causing many local samba artists involved in the city's carnival to lose their posts in the parades. Local arts directors and samba dancers were overlooked and lost their relevance due to the extensive hiring of carioca samba workers as a result of this new carnival model. Rita Maidana and other fellows who held prominent positions in the parades, opened new contacts to work in smaller carnivals in the Pampas region that had growing carnival parades, such as Bella Unión (Uruguay) and Alegrete (Brazil). In these places, Uruguaiana's carnival workforce was enthusiastically received due to the good reputation of the parades and the consolidated prestige of its carnival in the region.

The school of samba Emperadores de la Zona Sur from Artigas (Uruguay) hired Rita for the carnival arts direction in 2013 and 2014. The contract with the school of samba for that period included researching the plot, writing the synopsis, designing the costumes projects and floats of the Emperadores, and supervising the plastic-visual production work. To build up the project, Rita used her social network to indicate samba professionals for specific tasks in the carnivals she developed. She recommended a costume workshop in her neighborhood in Uruguaiana, and brought to Pampas region André Koppke, an artist who worked in the production of the floats in those years. André owned a costume workshop and provided services for schools of samba in Rio de Janeiro specialized in art painting, the final stage of making the props.

In Artigas, Rita and André introduced new techniques and notions of carnival organization that they had learned in other Brazilian carnivals, allowing their knowledge in the production of parades to be incorporated into carnivals and used in manufacturing techniques. Rita reused materials, reformulated carnival forms, and updated schools of samba in their contexts guided by the quality upgrade, based on her learnings in the Brazilian carnival.

André and Rita used production techniques based on the improvisation of real-time solutions such as those used in bricolage (Lévi-Strauss 2005). Bricolage is a type of knowledge based on the definition of the result to be achieved in the very act of finding the best ways to produce a certain object or effect. In bricolage, there is no prior definition of raw materials or utensils to produce a certain artifact. From a set of available materials and utensils, with wide use of reuse and substitution of elements, the ways to make a certain thing appear in the same moment of the act of manufacturing.

The carnival producers who worked in the workshops and spaces of the Pampas schools of samba elaborated their carnival projects making use of improvisation and the heterogeneity of materials for the manufacture of carnival objects. They adjusted the scarcity limits (with lack of raw material and labor), and developed largely techniques that adapted to the confections considering the options of recycling and recovery of the limited set of available materials, which reflected the possibilities of doing simultaneously their particular work.

The bricolage in carnival enabled the possibilities of appropriations and interpretations about the story to be told, and the definition of the plastic-visual project of the school of samba, which dialogues directly with the defined narrative, in a very flexible and adaptable way. A school of samba can purchase most of its plastic elements originally produced for other carnivals, rearranged in the transposition of elements and projects in different carnival contexts, which surpass geographical and cultural barriers. The restoration of carnival stuffs, reuses, recycling, and reforms are part of the resignification of the objects.

The strategy widely used in the carnivals of the Pampas is the purchase of objects from one or more carnivals to be replicated in a new location, with the consequent reforms and resignifications projected. This means a complete package that includes the storyline, the costumes, the allegorical elements, and the team specialized in the plastic arts production of the school of samba. Many times it can be included the hiring of plastic arts professionals, singers, couple of master room and flag holder and the drum director in the same negotiation.

Jéfferson Lima, a resident of Nova Friburgo in the mountainous region of the state of Rio de Janeiro and composer of sambas enredos (with victories at Imperatriz Leopoldinense, a traditional schools of samba), proposed to work with complete carnival projects ordered by countryside carnivals in Brazil. Jéfferson had experience in setting up carnival parades for the school of samba called Vilage in Nova Friburgo, where he participated in the artistic direction, purchasing lots of allegories, carnival materials, and costumes used in Rio de Janeiro's carnival.

Vilage school of samba worked on the objects used in the Marquês de Sapucaí carnival (Rio de Janeiro's carnival sambodrome), which were later rebuilt by its staff in Nova Friburgo. The artistic team recovered the pieces, modified a part of the costumes and, with the coordination of Jéfferson, made each piece fit the plot narrated in the avenue. Jéfferson calculated that the reuse of carnival objects, from different schools of samba in Rio de Janeiro, made Vilage spend up to four times less for each piece if compared to the manufacture of the same original object (considering the material used and the specialized labor employed). The economic advantage in looking for alternatives to self-made has reached great results. A school of samba that used the strategy of recycling materials was able to present parades with greater visual impact, increasing the quality of its parades and the chances to win its local competition, even with a very restricted amount of resources and sponsors.

Jéfferson Lima, when working in Uruguaiana, defined the same strategy of buying a complete package of carnival, from the plastic elements to the storyline, the samba music to be composed and indication of professionals for the arts project. Jéfferson developed his work in many schools of samba in Pampas region such as The Roux-inóis, the Bambas da Alegria, and the Unidos da Ilha do Marduque. His work was demanded by schools of samba from other cities in the region: Artigas, Santa Maria, São Borja, and Alegrete. Even without conditions to attend all the orders, Jéfferson intermediated the negotiations of carnival objects among schools of samba, espe-cially the resale of objects, which made long journeys along the variable life of these materials. Many carnival objects were reused more than once, being able to appear in countless parades in different geographically carnival competitions.

The rhythm of Jéfferson's production was determined by the conditions present in each creative process of producing a school of samba. The adaptations of the samba and the plot, the transformation of the synopsis and the modifications that followed the possibilities of buying used carnival objects determined the sequence of his work. There was no logical and temporal sequence established, the work was modified and adjusted according to the opportunities that he had found. By regulation, the synopsis of the plot and the organizational chart of the parade should only be given to the jurors on the eve of the parade. Thus, there was time for the arts director to transform the project according to the material conditions of carrying it out.

The achievement of material advantages focusing on the improved plastic-visual presentation stimulated the negotiations and the discursive arrangements to defend the purchase of complete carnivals in the routes between the Rio de Janeiro and Pampas carnival circuit. The economic and cultural circuits can be considered the most important factor to stimulate the translocal exchanges and contacts in the carnival into the border region. The greater standardization of the quality of the parade due to the reduction of plastic-visual differences justified the use of this strategy. Considering the visual projects, the discrepancy between the larger and smaller schools of samba could be counterbalanced in the competition through the purchase of materials for reuse, reform, or recycle, especially among those smaller groups with scarce funds and tiny social networks.

5   Hybridity and Globalization on Brazil's Southern Border

In recent years, the ways of financing carnivals on Brazil's southern border have brought the need for significant adaptations and changes in carnival management. Due to the country's political and economic crisis, some cities in the Pampas and other regions of Rio Grande do Sul have reduced their carnival budgets for samba schools or, in some cases, canceled their parades. In the year 2016 in Paso de Los Libres, the carnival groups parade did not happen because of a judicial dispute involving the official results of the previous year competition, added to the municipal fiscal crisis. The two most important groups, Carumbé and Zum Zum, had two years of preparation to return to the city's sambodrome only in 2017.

Budget reductions and carnival cancelations took place in several Brazilian states. The strong mobilizations among the samba followed the heated debate in the public spheres regarding the costs of the festivals, from the anticipated expenses to the forms of sponsorship. In Uruguaiana, the event had an important growth in municipal participation at the beginning of the century, becoming more expensive each year and with a greater need for public and private investments. It was known that the associations registered large debts that accumulated with each parade due to increased spendings, which indicated an unsustainable situation over the years, leading to recent changes in the management of financing and organization for the parade.

The most important theme about the planning of the carnival was the need to adjust the management of the event with long-term economic sustainability in mind. The strong dependence created on the need to hire external carnival labor, and the excessive purchase of materials and objects from other markets, brought the understanding that the carnival no longer developed the local economy as in other historical moments.

According to the samba people of the Pampas, the exclusively local production can bring an unequivocal loss in the aesthetic quality of the groups, and at the same time visibility and profits for the artistic producers of the region. On the other hand, the strategy of hiring professionals from Rio de Janeiro and reusing the objects increased the competitiveness, visual quality and, interest for the public that support the parades, as well as increasing the relevance of the Pampas carnival to other regions of their respective countries due to the attractions brought from other localities (Duarte 2016).

The carnivals of the Pampas region must be analyzed as a totality if we consider their potential of hybrid culture involved in the dynamic game of mixing between: the dimensions of the local context—such as improvisations, adaptations, and resignifications of their own—and the global rules and their arts forms—such as special elements, rules, regulations, and hybridity in each specific context (Hannerz 1997). We know that cultural exchanges, the transit of people and goods are not considered a recent fact in carnivals. If we consider the far-reaching anthropological theories, the population diasporas, the cultural loans, the movements of geographical displacements are phenomena correlated to the common existence of humanity since ancient times based on the flow of people and things through geographical spaces (Hall 2009).

What differs the most recent globalization from another one in distant times is that the global system, adjusted to the development of current capitalism, has consolidated new financial markets, with the growth of new creative cultural industries driven by information technology and the reduction of physical distance due to transportation development and advanced logistics infrastructure (Featherstone 1995). In this process, attempts at cultural homogenization, reinforced by common markets, are held back by the heterogeneity of local contexts and the claims of particular cultures: political struggles to establish structured identities in particular cultures. In Pampas' carnivals, the general characteristic is put forward links of negotiations between the local and global scales, thus emphasizing the circulation sustained by the translocality of this carnival culture.

Hybridity is another term for thinking cultural translation (Grímson 2011). It is not necessarily an appropriation or an adaptation. It is a process by which cultures demand

a revision of their own systems of reference, norms, and values, by distancing themselves from their usual or inherent rules of transformation. Ambivalence and antag-
onism are placed in each act of translation. Bhabha (2010) claims the malleability,
the dialogic and the dynamic work of cultural production on that terms. New positions, negotiations of meaning and representation of reality as "(…) a new place of
cultural enunciation, indeterminate and ambivalent by nature, since the conditions of
language as a structure of meaning, the positions of the subject and the implications
inherent in the enunciation, the reference" (p. 26).

The phenomenon of schools of samba in the triple frontier between Brazil,
Uruguay, and Argentina proposes, through the actions of their participants and their
contractors to work in their events, point out ways to think about culture and identification strategies in the updating of concepts, solutions to difference, hybridity, resignification, constant negotiations, and not fixable notions of cultures. The strategy
managed by the Pampas carnival for its growth, especially in the period after 2005
with the establishment of the extended calendar that covers several cities and their
parades, leads us to think about the festivals from the global scale constantly updated
in their local environments in the circuits of exchange (Appadurai 1997), such as
those experienced by Rita Maidana and Jéfferson Lima in the cities of Uruguaiana
and Artigas, as we presented previously.

The unstable balance presented in the translocal negotiations depends on the situations of intersubjective exchanges, on the interactions, on the cultural clash carried
out according to the historical conjuncture and the lived experience. In each new
carnival, there is an update of the political disputes and its strategies of realization,
the changes in the economic management of the festival and the practical senses of
the schools of samba around the material exchanges and attraction of new participants
and samba professionals.

Analyzing the socio-cultural practices in the carnival context, we have the possibility to investigate the dynamics of the carnival cultures in movement and the
Brazilian socio-cultural references closed to the political borders and beyond its
geographic limits. Therefore, the boundaries of carnival culture were not unique,
clear, fixed. They were always designed from the complex intersubjective experiences in the negotiations that flourished from the carnivals in the south Brazilian
border.

References

Appadurai A (1997) Modernity at large: cultural dimensions of globalization, 3rd edn. University
    of Minnesota Press, Minnesota
Bhabha H (2010) O Local da Cultura. Belo Horizonte (ed). UFMG
Canclini NG (2008) Culturas Híbridas: estratégias para entrar e sair da modernidade. Ed. da
    Universidade de São Paulo, São Paulo
Cavalcanti MLVC (1995) Carnaval Carioca: dos bastidores ao desfile. Ed. da Ufrj, Rio de Janeiro
Comaroff JJ (2003) Ethnography on an awkward scale: postcolonial anthropology and the violence
    of abstraction. Ethnography 4:147–179

Cunha FL (2006) Da Marginalidade ao Estrelato: o samba na construção da nacionalidade. Annablume, São Paulo

Duarte UC (2016) Carnavais Além das Fronteiras: circuitos carnavalescos e relações interculturais em Escolas de Samba no Rio de Janeiro, nos Pampas e em Londres. Unpublished Doctroral Thesis on Social Anthropology – PPGAS/UFRGS. Porto Alegre

Duarte UC (2016) Escolas de Samba nos Pampas: textos e contextos da interculturalidade no carnaval de Uruguaiana, vol 1. Revista Latino-Americana de Estudos em Cultura, Pragmatizes, pp 30–45

Featherstone M (1995) Cultura de Consumo e Pós-Modernismo. Studio Nobel, São Paulo

Grímson A (2011) Los Límites de La Cultura: crítica de las teorías de la identidad. Siglo Veinteuno Editores, Buenos Aires

Hall S (2009) Da Diáspora: identidades e mediações culturais. Ed. UFMG, Belo Horizonte

Hannerz U (1997) Fluxos, Fronteiras, Híbridos: palavras-chave da antropologia transacional, vol 3, no 1. Revista Mana. Rio de Janeiro, pp 7–39

Lévi-Strauss C (2005) A Ciência do Concreto. In: O Pensamento Selvagem. Campinas. Editora Papirus, São Paulo

Turner V (1974) O Processo Ritual. Ed. Vozes, Petrópolis

Igor Sorriso: A Narrative About Experiences and Evolution of Life Through Carnival

Igor da Silva Moreira, Rafaela Sales Goulart, and Fabiana Lopes da Cunha

Abstract This text brings an autobiographical narrative of the great interpreter, Carnival, and composer Igor Sorriso. The author tells us about his life, his discovery as a musician and lover of samba, and how his experiences were in various places in Brazil, and especially, as a samba interpreter in the great samba schools of Rio de Janeiro, such as Vila Isabel and, of São Paulo, in Mocidade Alegre. It is important to emphasize that, among his professional autobiography, Igor reveals how much he learned from his career related to Carnival: his perception about caring for his own voice, his contacts, and identifications that made him move to São Paulo, and finally, the recognition that the association/pavilion is the main flag of the Carnival, and therefore, its main patrimony.

Keywords Autobiography · Schools of samba · Igor Sorriso-Interpreter of Samba · Rio de Janeiro · Sao Paulo · Brazil

1 Presentation of Igor Sorriso

"I was introduced to the *Samba-enredo* in the Carnival at a very young age. I already had a certain proximity to this universe, because my father always liked it very much and I experienced the Carnival together with him. I remember that one day I was at a party, it was

I. da Silva Moreira (✉)
Brazilian Samba-Enredo Interpreter, "Igor Sorriso", Mocidade Alegre Samba School, São Paulo, Brazil
e-mail: igor87sorriso@gmail.com

R. S. Goulart · F. L. da Cunha
State University of São Paulo "Julio de Mesquita Filho" (UNESP), São Paulo, Brazil
e-mail: rafa_historia@hotmail.com

F. L. da Cunha
e-mail: fabiana.cunha@unesp.br

R. S. Goulart
Paulista State University Júlio de Mesquita Filho (UNESP), Coordination of Improvement of Higher Level Personnel (Capes), São Paulo, Brazil

F. Lopes da Cunha and J. Rabassa (eds.), *Festivals and Heritage in Latin America*,
The Latin American Studies Book Series,
https://doi.org/10.1007/978-3-030-67985-9_5

Fig. 1 The *cavaquinho* is a string musical instrument (4) widely used in Brazil's popular folklore

a June party (translator's note: The June festivals in Brazil are, in their essence, multicultural, although the format with which we know them today originated in the festivals of popular saints in Portugal: the Feast of St. Anthony, the Feast of St. John, and the Feast of St. Peter and St. Paul. The music and instruments used are: *cavaquinho* (Fig. 1), accordion (Fig. 2), *reco-reco* (Fig. 3), triangle or *ferrinhos* (Fig. 4), *zabumba* (Fig. 4), etc., and are at the base of popular and folkloric Portuguese music that was brought to Brazil by the people and immigrants from our mother country. The *caipira* clothes are a reference to the country people that populated mainly the northeastern region of Brazil, and one can find similarities in the way of dressing *caipira* in Brazil and Portugal) and a man who lived in our street, a childhood friend of my mother, called me and said: "let's go there with me".

He took me to the samba court of São Clemente (Translator's note: São Clemente is a Brazilian samba school in the city of Rio de Janeiro founded in 1961. Over the years, this school has been noted for its plots full of good humor and sarcastic criticism of the most diverse themes). And, on the way, when we were inside the taxicab, he gave me a tape recorder, and there was a samba-plot in it, which he had composed for São Clemente. His goal was to take him to the samba-plot contest in the eliminatory phase. The winning samba is that one which was sung during the Carnival parade, the following year. Then he told me: "I set up this samba and I want you to sing it, because I saw you singing around, I know you sing well, and I want you to sing this samba for me. But I said, "But what do you mean? Now?" [laughs] He said, "Yeah, you have 20 min to learn it. I said, "Oh, my God. Meanwhile, we went to São Clemente. When we got there, I sang the samba in a reasonable way, let's say so, and the samba qualified for the next week, for another phase of the samba

Fig. 2 Popularly known as accordion, it is a musical instrument composed of a bellows, free reeds, and two wooden harmonic boxes

Fig. 3 There are two types of reco-reco, one of them consisting of a bamboo or wood bud with transverse butchers and the other of a metal box and drumstick

Fig. 4 Percussion musical instruments, the triangle or little iron is made of iron or steel. The drum, popularly known in Brazil as "zabumba", is made of wood planks fixed with alternate shafts or metal

school qualifying. Unfortunately, after two weeks, when we presented it again, the samba was disqualified; but we created a bond of friendship and he called me to compose a samba together with him at Mocidade Unida de Santa Marta (United Youth of Santa Marta), which is in Dona Marta's community, in Botafogo, Rio de Janeiro, the first community that had the insertion of UPP, Police Unit. This was very chatted about throughout Brazil in Dona Marta's Hill. And I went there to Mocidade Unida de Santa Marta, it was my first samba composed, together with this friend, Walter, and this samba ended up becoming the winner. The school, at the time, was in access group C, the fourth group of Carnival in Rio de Janeiro.

An explanation is needed here: Group C of the samba schools in Rio de Janeiro is part of the fourth division of the Carnival schools in Rio de Janeiro. There are several parades of samba schools in Rio de Janeiro during the Carnival period. The most famous is the samba schools of the special group. In general, the school that loses this parade "falls" to the access group. Just as in football/soccer tournaments, like it happens in the Premier League in the UK or the "Bundes Liga" in Germany. Likewise, the winner of the access group "goes up" to the special group. In the case of the fourth division, the dispute for the title of Carnival champion takes place through assessments made by jurors divided into several questions previously stipulated by the league that organized the event. The champion is promoted to parade the following year in the third division.

The invention of the competitive parade of samba schools in Rio de Janeiro was shaped in 1932. As the number of registered schools increased, in 1952 the second division of Carnival was created. In 1960, the third division was so, and only in 1979, the fourth division of Carnival was formed. The name of Group C was changed a few times and in the first years its parade was organized by the Association of Samba Schools of the City of Rio de Janeiro (Associação das Escolas de Samba da Cidade do Rio de Janeiro—AESCRJ) until 2014, when the entity was extinguished. In 2015, the fourth division was renamed as Series C, being and commanded by the League of Samba Schools of Rio de Janeiro (Liga das Escolas de Samba do Rio de Janeiro— LIERJ), which also organized the second, third, and fifth divisions. To manage Groups C, D, and E in the Carnival of 2016, the "Cultural Association the Samba is Ours" (Associação Cultural o Samba é Nosso—ACSN) was founded, which lasted only one year. Starting in 2017, the Independent League of Samba Schools of Brazil (Liga Independente das Escolas de Samba do Brasil—LIESB) began to administer Series C, along with Series B, D, and E. But in a new change generated by LIESB, the fourth division of Rio's Carnival became the Access Group of Intendente Magalhães (former Series D).

After this parenthesis, we may recommence our history. Although samba was the champion, the school had no appointed singer yet. In groups of samba schools that are in the third, fourth or fifth division, they have great difficulty in getting components who would help with the music, props, and the organization of the parade. That was when the president of the school called me, right after the final that the samba won and he said: "Do you want to be the singer of our school? I was, at the time, only 16 years old. Then I said: "All right, let's go". I accepted the challenge. I became the singer of Mocidade Unida de Santa Marta (United Youth of Santa Marta), from group C. This was my first Carnival as a samba singer. I did not have a good debut, the group was demoted to the access group D [laughs]. That was very sad, but not enough to make me give up my career.

During that year's Carnival process, I ended up making many friends and a boy from Santa Marta Hill told me: "gee, How old are you, my son? I said: "I am 16". He said: "Well, you can still sing. You can sing in the myriad schools". I didn't know at the time. I said, "But what do you mean? And he said, "Yeah, there are countless schools." Then he took me to the "mirim school" of Salgueiro, which is called "Aprendizes do Salgueiro" (Salgueiro's Apprentices). It is also a very nice project in Rio de Janeiro, which is the "mirins" schools, and many *sambistas*, many people who are there in the Carnival, who work with Carnival, who live from Carnival, came from the "mirins" schools. I have many friends, who today are great carnivalists, great master loungers, flag-bearers, interpreters who started in the myriad schools. And I was one of those, I went there to Aprendizes do Salgueiro, and when I got there I had a bunch of boys singing and I, being shy, picked up the last microphone. And I learned to sing at Aprendizes do Salgueiro, where I stayed for 3 years, the age limit is up to 18 years, so I sang in three Carnivals at the samba school Aprendizes do Salgueiro, as a support singer. In this period, I reconciled my participation in the mirim school with the United Youth of Santa Marta, as an official interpreter, learning, evolving, improving myself as a musician and interpreter.

It is important to point out that Santa Marta has a very strong connection with São Clemente, which was the first samba school in the southern zone of Rio de Janeiro. And then the drummers from Unidos de Santa Marta took me to São Clemente one day, for a rehearsal. There, they talked to the president and asked me to sing a samba on stage. And then I sang the samba and the president of the samba school liked me and called me to start performing at the school. He said: "Do you want to be part of St. Clement?" I said: "Fine! We can see everything". And then I became one of Saint Clement's support singers. At the time, in 2006, São Clemente was in the access group A, the second group in Rio de Janeiro. In 2007, São Clemente was the champion of the access group, which made it possible for me to make my debut in 2008 in Rio de Janeiro, singing in the special group, but still as a support singer.

At the end of 2009, in December, more precisely, the singer of St. Clement, Leonardo Bessa, left school, and when this happened, there were only 2 months left to Carnival. At Tuesday's rehearsal, the president went on stage and spoke into the microphone: "Leonardo Bessa is no longer part of our school and our official singer is now Igor Sorriso. [laughs] Wow, I was upset! Imagine the cold in my stomach! It's a big responsibility to be the singer, the official interpreter of a samba school and I always kept looking at the guys, my idols, Neguinho da Beija-Flor, Neguinho da Estácio, or Jamelão, and I kept imagining what it was like to sing as an official singer, to pull the team from the school. That's why I like to be the samba "puxador enredo" (the official interpreter of the school's samba), because he is very representative of the school. The term "puxador" is very demonstrative! I like this term, "puxador", because we bring emotion and joy to a lot of people with our presentation. We bring a lot of people along with us; there are people who are there at the end of the school, at the end of the wings, and who cannot see what is happening elsewhere in the front of the school. And there, through our voice, through what we are encouraging, what we are saying, that the person enters the avenue and manages to make a good evolution and help the school performance.

Continuing with the story, I became the official singer of São Clemente. I had two months to prepare myself, to develop a good work as an official singer in the Carnival of Rio, in the access group. And in February 2010, I debuted on the avenue as the official singer. São Clemente was the champion of the access group and went up to the special group of the samba schools in Rio de Janeiro.

Now I wanted to talk a little about my work as a samba school singer, because it is a very difficult process, a little ungrateful to deal with, because it is a pathway in which we do not have much support or guidance. I learned a lot in this art of singing samba storyline, mostly observing the singers, of "achisms", of living... I never had a professional orientation. Thus, this is the popular music itself, isn't it? In the literal sense of the word, because we learn here and there, to perform in the best possible way, because it is a hard competition. The school receives grades for its performance, and you are there, within this competition, and your performance is being thoroughly evaluated. When you have a good performance, you stand out, you value yourself and you are valued by the jury members. But when you do not perform well, you get stoned, you get massacred and this often happens without having any serious foundation, without having any musical knowledge.

As I did well, the school renewed my contract, and I was the official singer of the Carnival 2011, São Clemente, the samba school of the special group. That was when I tried to get deeper into this musical issue. Before that, I thought I sang well, just because I liked to sing. Therefore, I started to attend singing classes, and I found out that I did not sing well at all. But, it was very nice. And, since that period, until today, I have a greater concern with my voice care, with my proper rest, with my adequate food, with my regular performance in rehearsals, and all this made me rapidly evolve a lot. Sometimes I wanted an authorized opinion, I wanted a greater musical basis, I wanted a firm support, but in samba schools, sometimes we do not get a deeper support in relation to this, to our art. Therefore, I met my singing teacher and stayed with him for about four or five years. Soon after, I also started the phonoaudiological work, seeking the appropriate recovery of the voice, the quality of my voice, to keep the voice always with relevant quality, because the samba plot is one of the most aggressive styles for the human voice. This is because, many times we sing without adequate conditions, sometimes with the accompaniment of very precarious sound equipment, but with 200–250 rhythmist drums. This is an absurd sound competition, and it will wear out the voice if you do not take proper care of it, if you do not hydrate yourself as needed, if you do not rest enough. As time goes by, your voice will no longer be the same, so I usually say that... People even play, don't they? But Jamelão drank, turned the night. Then I answered: "How many Jamelões exist? Only one!" A lot of people from such time who had a lot of exaggeration in their voices and who did not have a follow-up, who did not have professional support, could not take their career further on. Thus, it was very important that when São Clemente debuted, I debuted in São Clemente, in the special group in Rio de Janeiro, in 2011, and São Clemente stayed in this group, right? São Clemente stayed, and I had the happiness of receiving the golden banner, the revelation of Carnival and then the doors of Carnival were finally opening for me.

It was after that that I met the Carnival of São Paulo, and also the Carnival of Porto Alegre, the Carnival of Uruguaiana. That is because the Carnival of Rio de Janeiro is the biggest Carnival in Brazil, speaking in media terms, and therefore, it gives you an international visibility. Thus, through the Carnival of Rio de Janeiro, I made many friendships, I made many contacts, and met other Carnivals. And I had no notion of all these parties: I did not know the town of Uruguaiana, I had never heard of it, I did not know that Brasilia, the capital city, had a Carnival, nor that there were parades in Manaus, Belém or in Vitória do Espírito Santo. I happen to know the various parades and associations of different cities and I made friends and contacts and met many people in these Carnivals. In my Carnival, in different cities, I made many friendships, because the Carnival unites, the Carnival is aggregating, there is no way. I met many people, I made many contacts and today I live in and of Carnival.

I stayed in São Clemente until the Carnival of 2015, that is, six years as an official singer. It was there that I received an invitation from Unidos da Vila Isabel (United of Vila Isabel) to be the official singer of the 2016 Carnival. And I went to venture into the land of Noel Rosa, from Martinho da Vila. I stayed in this samba school for three years. Unidos da Vila Isabel is a passionate, vibrant community, it brought me back a lot of things and a lot of learning too. But before that, still in 2012, I started to parade in the São Paulo Carnival.

My first contact with the Carnival of São Paulo was as a support singer at Mocidade Alegre and in the following Carnival I went to Acadêmicos do Tucuruvi (Tucuruvi Academics) to be the official singer of the Carnival of São Paulo. I returned to Mocidade Alegre, in 2014, also as an official singer, and there I stayed for two more years as a Mocidade Alegre singer. Then I gave up this double journey, because it was very tiring, very complicated, rehearsing in Rio de Janeiro, catching the plane, going back to São Paulo, returning to São Paulo, every week, two or three times a month, was really tiring. Therefore, I chose to sing only in Rio de Janeiro, starting in the Carnival of 2017.

However, there is one thing that I admire a lot, which today has become much more complicated, because Carnival has become much more commercial. But, there is something special, that I admire, which is the identification with a samba school. I, even far away, created a very big identification with Mocidade Alegre, which is the school I'm working today. So, I stayed in Vila Isabel, until 2018, and in the Carnival of 2018, I made a very difficult choice. Let's say it is the opposite of what people usually do, because the singers from São Paulo all want to be in Rio de Janeiro; and I did the opposite way, I left Rio de Janeiro's Carnival, by my own will, and decided to go back to São Paulo to sing only at Mocidade Alegre.

Thus, my priority today is the São Paulo Carnival, today I am the official singer at Mocidade Alegre. Along with that, I moved: I left Rio de Janeiro and I have been living in São Paulo for one year, exactly one year ago I took my whole family and I am living in Mocidade Alegre district. My identification and performance at Mocidade Alegre is much broader.

Today I am more active in school, I am more involved with the projects, I have more time to dedicate to Mocidade Alegre. I left the position in Rio de Janeiro. And what I heard from many people is: "Are you crazy? How are you going to leave Rio de Janeiro's Carnival to go to São Paulo?" which in theory is the second Carnival. But there are things that are a priority in our lives. Today my priority is to be able to do, to create an identification, to be able to stay a long time in a samba school, and Mocidade Alegre has always made this possible for me, it has always given me support so I can work. With such a group, I got a better performance as a samba player, as a musician. Mocidade Alegre, for giving me better support, brought more quality to my work and my life. And all this came together with a great feeling of identity with the school group, with the Carnival association.

In the two years that I stayed out of school, I received many messages along the social networks: "Well, you need to go back to our school. Reading and listening to this motivates us, makes us want to be back, to be together with this community. Today, I am part of the São Paulo Carnival and I am no longer in Rio de Janeiro Carnival. In the background, I am also in these other Carnivals.

This year, for example, I had the opportunity to sing at the Artigas Carnival in Uruguay, through the Uruguaiana Carnival. I have been singing for seven years at the Carnival of Uruguaiana. I have already created an identification with the Carnival of Rio Grande do Sul, because I sang during several consecutive Carnivals, and the Carnival of Uruguaiana is the most representative Carnival of the southern region. At the border with Argentina, there is the Paso de Los Libres Carnival. In the town of Artigas in Uruguay, there is also Carnival. Through Uruguaiana, I managed to create some contacts, some friendships there in the region. This year I sang in the Carnival of Artigas in Uruguay, a very nice experience. I was very happy to participate, and it is very different from Carnival here in Brazil. Here there is human warmth, people interaction. There in Artigas, everybody sits in the beach chair, watching the schools parading [laughs], and admiring with binoculars, with small flags. Everybody sits there. It is very nice to see the people with the beach chair on the street, watching the school parade.

I am a person who greatly respects this opportunity that was given to me. It is a very big responsibility to be the official singer. I confess that every year when I arrive at the avenue and play the siren to start the parade, it gives me a cold in the absurd belly, I get very nervous, my throat gets dry [laughter]. But we go and start singing. When we do it, it has gone, it is over. I am very happy to be able to represent Carnival, to be the spokesperson for a community, which is very important. Since we are here talking about patrimony, I wanted to end by saying that the most important patrimony of an association is its flag, its pavilion. So, regardless of personal issues... because they will always have problems, they will always have discussions, in their own work... But today, I respect all the pavilions, all the flags, all the schools that I have represented, and I try to pass this through my singing, my voice, the seriousness of my work, and I always try to seek, to give my best, to be able to create this identification with the school. To understand the history of the school, to understand everything that has already happened, everything that has been done, and to pass my maximum so that I can spend a lot of time, behind the song, behind the identification, to continue doing one of the things that I love most in life, which is to sing "*Samba-enredo*".

Thank you for your attention, for the affection, for the attention and this is my story [applause].

Return of Igor Sorriso, Mocidade Alegre—2019.

In: https://youtu.be/f9heQNAr3BE.

Religious Celebrations and Its Safeguard Public Policies in Brazil: Directions

Rafaela Sales Goulart and Fabiana Lopes da Cunha

Abstract Based on the explanation of the historical trajectory of the heritage concept, and more specifically, its appropriation as a political object of identity projections, this manuscript will reflect on the unfolding of cultural heritage in Brazil, emphasizing the field of intangible heritage. In this field, in turn, it is found the category "celebrations", created as a result of the instrumentalization process of public safeguard policies for such cultural goods (Decree No. 3,551, of August 4th, 2000). In this context, we aim to discuss the main limits and possibilities that encompass studies, registration processes, and social actions that favor the safeguard and continuity of these celebrations, especially the religious ones. The directions have been chosen and will be shared not only because of our research experience with groups of *folia de reis* from the interior of the state of São Paulo, but also because of the challenges inherent in instrumentalizing public policies for such celebrations.

Keywords History · Memory · Intangible Heritage · *Folia de Reis* · Public Policy · Brazil

1 Patrimony's Trajectories: A Brief Contextualization

The concept of patrimony (*patrimonium*) has a Latin origin, referring to everything that belonged to the father/father of a family (*pater/pater familias*) and what was inherited as private property, within the law of Roman aristocratic societies. However, with the rise of Christianity and the medieval Church, which, above all, from the sixth century onwards inaugurated a kind of cult of saints and relics, the heritage, characterized by the idea of private and individual property, which belonged only to the

R. S. Goulart (✉)
Paulista State University Júlio de Mesquita Filho (UNESP), Researcher financed Coordination for the Improvement of Higher Level Personnel (CAPES), Assis, São Paulo, Brazil
e-mail: rafa_historia@hotmail.com

F. L. da Cunha
Universidade Estadual Paulista "Júlio de Mesquita Filho" (UNESP), Sao Paulo, Brazil
e-mail: fabiana.cunha@unesp.br

elites, gains a collective value to the extent that religion and faith can be socially shared symbolic elements. With the historical movement of the Renaissance and its consequent humanism, antique dealers were created to rescue and protect works and objects that referred to classical antiquity. Thus, although humanists rejected medieval times, they reinforced with the antiquarians the idea of heritage as something linked to a common past and that, for its relevance, should be preserved (Funari and Pelegrini 2006, pp. 10–13).

Interestingly, the conception of heritage that we analyze is linked to a western historical trajectory, being extremely relevant to think now about the fall of the old monarchic regimes and the invention of European national states from the eighteenth century. According to Funari and Pelegrini (2006, pp. 17–19), it is in this modern context that heritage, within the tradition of Roman law, is shaped as something public, belonging to the nation-state. It means that the states have the right to possession and protection of what was now considered historical and cultural heritage, located on public or private properties.

The idea of inventing the Nation-State links to the production of a sense of national identity produced through symbols common to a society: monuments, anthems, flags, language, etc. Identified with these symbols, populations could feel that they belong to their nationalities, which would serve as a relevant element of a social union in times of crisis, and consequently, of disunity, when it is understood that what is different can also be the common enemy of a nation. The world wars of the twentieth century and their overwhelming results are examples of how nationalism has worked.

It is essential to reflect on the notion of value that constitutes the heritage of an individual or society (Fonseca 2017, p. 33). Heritage can be something of spiritual value that we receive from our ancestors, objects, or teachings that we could only mention individually. Within this logic, a patrimony common to society would have to possess and give rise to a collective feeling of recognition, as in the case of language. However, when we resume the expression "invention of nations", we want to indicate the idea that a small part of society conceived and induced the belief of what is a nation, and therefore, of what is its official patrimony. At this point, it is important to differentiate "tradition" from "custom":

> The purpose and characteristic of "traditions", including those invented, is the invariability. The real or forged past to which they refer imposes fixed (usually formalized) practices, such as repetition. The "custom", in traditional societies has the double function of the engine and steering wheel. It does not prevent innovations and can change to some extent, although it is evidently hindered by the prior sanction, historical continuity, and natural rights, as expressed in history. The real or forged past to which they refer imposes fixed (usually formalized) practices such as repetition. (Hobsbawm and Ranger 2018, p. 9).

Hobsbawm and Ranger make us understand better that those considered national assets do not necessarily invoke the sense of identification and value to certain social groups that make up the so-called nation. And it is precisely this reflection that we consider essential to give continuity to this text, since it raises problems such as: how many languages and dialects had, for example, what we understand today as Brazil when the Portuguese throne literally occupied this territory in the nineteenth century? How many social groups, customs, and identities have been erased from this territory since the arrival of the first Europeans in the sixteenth century, and since the nineteenth century, have been silenced by the imperial government in the name of our nation's invention? How did the construction of national identity take place in the Brazilian Republic?

We certainly do not intend to answer all the questions raised, but to induce a critical reading of the complex history of heritage.

Before entering properly into the disputes of memory around the cultural heritage, it is still necessary to emphasize the period from 1914 to 1945, which marked the beginning and the end of the two great world wars. This moment was characterized by nationalism and imperialism based on racist and fascist ideas of nations' superiority and populations over others. In this way, all the material, human, and immaterial post-war destruction resulted in the condemnation of nations that committed acts such as genocide, as in the case of Germany, and of countries that, even considered democratic, continued to develop racist policies over their colonies throughout the twentieth century, as in the case of England (Funari and Pelegrini 2006, pp. 21–22).

In this respect, the social unrest, economic, and political wars promoted the constitution of entities such as the United Nations Organization (UN) and the United Nations Educational, Scientific and Cultural Organization (UNESCO) created in 1945. The UNESCO aimed, as its name presumes, the alliance of nations based on the idea of promoting peace and defending human rights and social, cultural, and environmental diversity, which comprise the most varied places and populations of the world. The notion of preservation, for example, has been adhered to in order to safeguard and economically enhance cultural varieties, which in the case of cultural heritage, would be possible through historical and cultural tourism. Besides, such activities linked to these heritages would function as a symbolic mechanism to repair the memories erased during the massacres in the world wars. The Auschwitz-Birkenau site, for example, was recognized as a World Heritage Site by UNESCO in 1979.

Funari and Pelegrini (2006, p. 27) presented that the universality surrounding the idea of the protection of UNESCO World Heritage Sites does not, however, translate into official recognition of World Heritage Sites of and by all nations. According to the authors, Europe and North America are the territories where the most significant number of identified heritages are found, which may represent mental, cultural, and political backwardness in relation not only to the understanding of what the world heritage would be, but also to the intentions and social and memory conflicts that involve their recognition. Not by chance, the history of the concept of cultural and environmental heritage needs to be discussed, especially given the various immaterialities surrounding them (Funari and Pelegrini 2006 pp. 17–19); after all, they encompass the political and economic interests of nations over others.

Directing the magnifying glass to Brazil, and more specifically, to the federal policies of preservation of cultural heritage in the country, Fonseca (2017) indicated the 1930's as a landmark for the elaboration of policies to protect what would be the national heritage of the country and for the very creation of specific organs of identification and choice of its patrimonies, which should remit a historical and identity value to the Brazilian people. In this sense, under the direction of Rodrigo

Melo Franco de Andrade, the Institute of National Historical and Artistic Heritage (IPHAN) was created in 1937 to fulfill this function. Article 1, of Decree-Law Nº 25/November 30, 1937, organized the protection of the national historical and artistic heritage and defined it:

> The national historical and artistic heritage is the set of movable and immovable assets existing in the country and whose conservation is of public interest, either because of its attachment to memorable facts in Brazil's history or because of its exceptional archaeological or ethnographic, bibliographic, or artistic value.

In the description, "movable property" refers to what is identified today in Article 216 of the Brazilian Federal Constitution as material assets which include monuments, objects, and constructions. The "real estate", on the other hand, would be the current immaterial heritages; that is, knowledge and doing, as rituals and oral expressions, that the human being apprehends with his peers and shares from generation to generation (Brazil 1988).

Despite the classifications, it is important to say that during the period of creation of IPHAN, the "memorable facts of the history of Brazil", and therefore, the symbols of the Brazilian national identity, were specifically guided by the aesthetics of the works and their exceptional character. Another factor to be highlighted is that the selection of movable cultural assets was strengthened by institutionalizing the teaching of architecture and engineering in Brazil, through the School of Fine Arts and the Polytechnic Course, which formed part of IPHAN's workers and technical experts (Fonseca 2017). At that time, for example, Ouro Preto was listed as an architectural and urbanistic group (Process Number: 70-T-1938; Book of Tombo Fine Arts: registered on 04/1938). In 1980, the city was recognized by UNESCO as a World Heritage Site (Tombamento Process: 64-T-38; Registration Number: 124; Registration Date: 05/09/1980).

Returning to the excerpt, the word "conservation" denotes more problematic about the criteria of selection and treatment of the Brazilian cultural heritage. If conservation is to preserve in the original way, how to conserve immovable/immaterial assets? Moreover, even if only movable/material assets were conserved, as occurred in the history of IPHAN until 2002 (IPHAN 2018, p. 12), one of the points we must reflect on is that said conservation should be done to assets of "public interest". However, it is known that the public until then: economic and political elites did not represent, and still do not today, the ethnic and cultural multiplicity that make up Brazilian society and history.

Given the changes in the context of the second half of the twentieth century, the history of heritage in Brazil, started moving towards the democratic bias, through initiatives, like:

- 1970–71: Holding of the Meeting of Governors and writing of the Letters of Commitment of *Brasília* and *Salvador*, which unfolded actions of joint preservation between cities, states, and federation, and therefore, allowed the decentralization of heritage policies;
- 1973: Creation of the Historic Cities Program (PCH), aiming to revitalize monuments, create infrastructure, and develop tourist activities in needy regions;

- 1975: Constitution of the National Center of Cultural Reference (CNRC) from the leadership of Aloísio Magalhães, which made possible the creation of the concept of cultural reference/good, appropriate later in the 1988 Constitution;
- 1979: Merger of Sphan (National Historical and Artistic Heritage Service—which had a certain prestige and technical competence), PCH (which had financial resources), and CNRC (which had a modern and renewing vision of heritage); and creation of the National Pro-Memory Foundation (FNpM). The latter, in turn, would fill gaps between institutions and deconstruct dichotomies between erudite and popular culture;
- 1981: Creation of the Secretariat of Culture (SEC) of the MEC and elaboration of the Guidelines for the operationalization of the MEC cultural policy;
- 1985: Creation of the Ministry of Culture (Minc);
- 1986: Establishment of the *Sarney* Act, which enabled fiscal incentives and public–private partnerships for the promotion of culture;
- 1988: Publication of the new Brazilian Constitution with official recognition of material and immaterial cultural heritage (Article 216), covering "manifestations of popular, indigenous and Afro-Brazilian cultures, and other groups participating in the national civilization process" (Article 215, item 1); such recognition is prescribed in the document not only by the public power, but in partnership with the communities holding the cultural assets (Article 216, item 1);
- 1991: Establishment of the Tax Incentives Law (Law N° 8.313);
- 1997: Holding of the Seminar "Intangible Heritage: Strategies and Forms of Protection", in *Fortaleza*; and writing of the Charter of *Fortaleza*, which attested to the commitment to elaborate guidelines and create legal and administrative instruments for the patrimonialization of intangible cultural assets.

As you can see, we have tried so far to locate Brazil in the historical panorama of political appropriations of the concept of heritage. With an emphasis on the twentieth century, we locate the 1930s, when IPHAN was created, as a troubled phase worldwide. We sought an identity that would insert the South American country in the global scenario, and in this sense, many memories, customs, and facts of this history were silenced as a result of the Western patrimonial logic.

In Brazil, until the end of the twentieth century, we could say that the majority of our patrimonies were recognized, which referred to the history of domination of Europeans over the new world and of economic and political-dictatorial regimes, which characterizes the appreciation of an official memory. According to Pollak (1989), the official collective memory is represented by the state through partisan ideologies and by institutions of power that seek its broad and effective legitimization. Thus, such "memory framed" in historical and cultural heritage would be the personification of certain traditions as a way of keeping them alive in memory, acting in defense of elite groups, and thus establishing a national memory through certain social coercion.

On the other hand, we stress the relevance of social movements at the end of the century. The struggles of the black movement in Brazil, for example, have contributed to advances in the understanding of the concept of heritage and its institutional applicability. The recognition by IPHAN of the White House Terreiro from *Engenho Velho* (*Ilé Axé Iya Nassô Oká*) from *Salvador* (*Bahia*) in 1986, is a clear example of this achievement. According to Funari and Pelegrini (2006, p. 50), the singularity of this tumbling is due to two reasons: first, because it is a space that comprises a religion with an African matrix; second, because the tumbling includes not only the mobile/material constructions of the terreiro, but also the trees and sacred objects that substantiate the senses of the religious ritual of *Candomblé*.

However, it is important to mention that the innovations experienced in Brazil, were the result of a changing international context. The 1985 Declaration of Mexico and the Recommendation on Safeguarding Culture and Popular Tradition promoted by UNESCO in 1989, are examples of the concern with "[…] ways of living, knowing and doing of traditional communities, considered "treasures of humanity" (Rovai 2016, p. 241).

Therefore, it is reflected that immateriality/intangibility involving physical constructions are important elements to be considered in the processes of recognition of cultural goods, even if they fit into the mobile/material category (Menezes 2012). In the case of the White House Terreiro, its tumbling also represents the remnants of a slave Brazil, that is, a national past where economic, political, and cultural elites, if "winners", were only achieved through hard work and reinvention by the so-called minorities (Indians, African slaves, poor immigrants, etc.), which have historically been silenced.

All this represents a long heritage trajectory, which is still constituted through memories and social identities in disputes.

2   Immaterial Heritage Policies in Brazil

The trajectories of the concept of patrimony and the conflicts of memory inherent to this history have unfolded in Brazil, specific policies to safeguard intangible heritage. The Charter of *Fortaleza* (1997) can be identified as a precursor document of a compelling call to an organization, study, and planning of actions that would be promoted through the partnership between public agencies, university institutions, and several entities linked to society. This document resulted in Decree N° 3,551 of August 4th, 2000, which instituted the registration of intangible assets that constitute the Brazilian cultural heritage and created the National Program of Intangible Heritage (PNPI).

To facilitate the registration, 4 books were established (Article 1, item 1). Namely:

I - Book of Knowledge Registration, where knowledge and ways of doing rooted in the daily life of the communities will be registered;

II - Book of Registration of Celebrations, where rituals and parties that mark the collective experience of work, religiosity, entertainment, and other practices of social life will be registered;

III - Book of Registration of Cultural Expressions, where literary, musical, plastic, scenic, and playful manifestations will be registered;

IV - Book of Registration of Places, where will be registered markets, fairs, sanctuaries, squares, and other spaces where they are concentrated and reproduce collective cultural practices.

The detailed description of what could be registered as a cultural reference of Brazil, inaugurated in the new century, the concern to elucidate and protect democratically, memories, and social identities until then silenced in the history of cultural heritage policies. To this end, Article 2 of the Decree details which could initiate the registration process: ministers and institutions linked to the Ministry of Culture, State, Municipal and Federal District Secretariats, and civil society or associations. Thus, after the enactment of the 1988 Constitution, known as the Citizen Constitution, participatory public policies were constituted, which, in the case of heritage, gave space to the social groups themselves to resort to official recognition of their cultures.

Article 15 of the Convention for the Safeguarding of the Intangible Heritage from October 17th, 2003, shows the idea of shared management (UNESCO, 2003), which fits in with what we refer to as "participatory public policies". Shared management" is the concept:

> [...] a management model that, as opposed to the centralized management model, is carried out jointly by different actors, agencies and institutions in order to achieve common goals and objectives, based on cooperation strategies and the engagement of different entities in decision-making processes, action planning, problem-solving, analysis and evaluation of results. (IPHAN 2006, p. 23).

Continuing this major step, on April 12th, 2006, Decree Nº 5,753 was issued in Brazil, sanctioning the Convention for Safeguarding the Intangible Heritage, adopted on October 17th, 2003, in Paris (France). It means that the State assumed the responsibility to create safeguard measures for its cultural assets before an international organ such as UNESCO.

In the 2003 Convention, UNESCO already announced its concern, above all, with intangible heritage belonging to indigenous communities, highlighting the intolerance and economic-territorial expansion present in a globalized society, which could lead to the deterioration and disappearance of cultural manifestations of these historically marginalized groups. Not by chance, the *Yaokwa* Ritual of the indigenous people *Enawenê Nawê* was included in 2011, the year after obtaining the title of intangible heritage of Brazil, on the List of Intangible Cultural Heritage Requiring Urgent Safeguard Measures by UNESCO. This cultural asset is guided by a cosmology that is based on the cycles of nature, which enable the organization of the collective fishing ritual.

According to Opinion Nº 015/10/CGIR/DIP/Iphan, for the maintenance of the indigenous ritual of the *Enawenê Nawê* it is necessary "1. The protection of the biodiversity that characterizes the region and, 2. The integrity of the logics that govern the systems of production and transmission of the knowledge and know-how associated with this Ritual". The importance of instrumentalizing and applying public heritage policies is evident, above all, because it enables the contact and social mobilization of holders of cultural assets in the face of protecting their heritage. On the other hand, a question arises: how can we guarantee the success of such a policy and the safeguarding of this reference in a place that, even if demarcated as indigenous land, is threatened in the face of economic development projects that provoke drastic changes in the environment?

Opinion n° 015/10/CGIR/DIP/Iphan itself accuses projects of constructing small hydroelectric plants, which could interfere not only with the fishing ritual, but also with the very subsistence of the ethnic group.

Although the questioning of the indigenous celebration suggests a greater reflection on the environmental future of Brazil and the world, we use it to direct our discussion on the complexities that are part of the process of registration, safeguarding, and sustainability of intangible heritage, which, integrated into the most varied historical-social dynamics, suggests conflicts and negotiations of various orders.

In the text "Challenges in the patrimonialization of intangible assets of a religious nature", Abreu and Magno (2017) proposed a debate that is more in line with what we intend here. Among the article's contributions, they pondered the difficulties of applying methodologies produced by public agencies such as IPHAN, especially when it comes to immaterial goods that are delineated by belief and devotion. Thus, they use a case study on the *Folias de Reis* of *Valença* (*Rio de Janeiro*) to reflect on the possible problems that the State University of Rio de Janeiro (UERJ) found when making the Inventory of the most varied folias of the kings of Rio de Janeiro. Among the problems cited by the authors are the variety of groups, and therefore, the identity characteristics of the celebration in a single city; and urban modernization, a phenomenon that can reconfigure the cities, while at the same time deconstructs the customs of the celebration.

Although these are distinct celebrations, with different beliefs, customs, and populations, both the indigenous ritual of the *Enawenê Nawê* and the *Folias de Reis of Rio de Janeiro* and the rest of Brazil, indicate the dilemmas to be explored when it comes to the application of policies for the registration of celebrations that engender religiosity and tradition.

Thus, by pointing out the institutional milestones of the instrumentalization of the policies for safeguarding celebrations in Brazil, we will highlight, in the next topic, the celebrations registered until 2019, and the limits and possibilities that are involved and still involve the process of safeguarding these intangible assets.

3 Limits and Possibilities for Safeguarding Brazilian Religious Celebrations

As indicated, Decree n° 3.551/2000 also instituted the PNPI, which should implement a specific policy of inventory, referencing, and valuation of cultural assets at the federal level (Article 8). The Program was regulated by IPHAN Ordinance n° 200, of May 18th, 2016, which reinforces the application of methodological instruments (National Inventory of Cultural References—INRC and National Inventory of Linguistic Diversity—INDL) that facilitate the research and documentation of cultural assets to be registered as intangible heritage of Brazil (IPHAN 2000).

Regarding the registration process, it has 3 stages: preliminary analysis, technical instruction, and final evaluation (IPHAN 2016). About the date and purposes of the implementation of the safeguard of the registered assets:

> It is expected to begin during the first decade after registration, with a view to strengthening the autonomy of the holders/producers of the cultural good in the production, reproduction, and management of its heritage and the sustainability of the cultural good in the medium and long term. (IPHAN 2015).

To accomplish this purpose, IPHAN acts within 4 axes: 1—social mobilization and policy reach; 2—participative management in the safeguarding process; 3—dissemination and valorization; 4—cultural production and reproduction (IPHAN 2015).

Between 2002 and 2019, 48 Brazilian intangible heritage sites were registered, 13 of them belonging to the Book of Knowledge, 13 to the Book of Celebrations, 18 to the Book of Forms of Expression, and 4 to the Book of Places. Of these cultural assets, we highlight in Table 1 the Celebrations and other information on dates, states, and areas of coverage.

One of the first observations to be made is that, except for the *Bumba-meu-Boi* Cultural Complex of *Maranhão* (5) and the Bumbá Cultural Complex of the Middle Amazon and Parintins (13), both of which are statewide, most of the celebrations registered by IPHAN are restricted to their own cities or, in the case of the *Yaokwa* Ritual, to the Enawenê Nawê Indian Land (3). This evidence signals the complex and delicate situation of the technical team responsible for the INRC, because regardless of the area of scope of the records, they need not only to raise documentation for the study, characterization of physical supports (clothing, flag, masks, instruments, etc.) and identification of technical skills present in the celebrations, but they also need to understand, from contact with the holders of the cultural asset, the meanings they attribute to each of the elements and festive moments, in a kind of translation of intangible heritage values. Hence, the relevance of acting along the lines set out above, for it is precisely in this process that the bonds of trust between the technicians of the Institute and other regional and local bodies are formed, with the community to benefit from the heritage policy.

A clearer example of the complexity of identifying the intangible in a single celebration/location and of applying public policies for its maintenance can be given from an experience report on research carried out between 2013 and 2016, whose theme was the *Folia de Reis* de *Florínea* (*São Paulo*). This work identified the symbolic union of two festive companies, initiated in the 1990s and consolidated by the creation of the "Folkloric Association of Kings *Flor do Vale* of *Florínea*", in 2013. The union of the groups took place precisely because of the historical reconfigurations of the city, in the regional context of mechanization of agricultural labor, and the arrival of agribusiness in the west of *São Paulo*, from the 1970s (Goulart 2018).

Table 1 Celebrations registered between 2002 and 2019 as Immaterial Patrimony of Brazil

Celebration	Date of registration	State	Area of coverage—city
1. *Círio de Nossa Senhora de Nazaré*	05/10/2004	*Pará*	local—*Belém*
2. Party of the *Divino Espírito Santo* of *Pirenópolis*	13/05/2010	*Goiás*	local—*Pirenópolis*
3. *Yaokwa* ritual of the *Enawenê Nawê* indigenous people	05/11/2010	*Mato Grosso*	local—Indian land *Enawene Nawe*
4. Party of the *Sant'Ana de Caicó*	10/12/2010	*Rio Grande do Norte*	local—*Caicó*
5. Cultural Complex of *Bumba-meu-Boi of Maranhão*	30/08/2011	*Maranhão*	State
6. Party of the *Divino Espírito Santo* of the City of *Paraty*	03/04/2013	*Rio de Janeiro*	local—*Paraty*
7. Party of the *Senhor Bom Jesus do Bonfim*	05/06/2013	*Bahia*	local—*Salvador*
8. *Glorioso São Sebastião* festivities in *Marajó* region	27/11/2013	*Pará*	local—region of *Marajó*
10. Party of the *Pau de Santo Antônio de Barbalha*	17/09/2015	*Ceará*	local—*Barbalha*
11. *Romaria de Carros de Boi da Festa do Divino Pai Eterno* of *Trindade*	15/09/2016	*Goiás*	local—*Trindade*
12. *Procissão do Senhor Jesus dos Passos* of *Florianópolis*	20/09/2018	*Santa Catarina*	local—*Florianópolis*
13. *Boi Bumbá do Médio Amazonas* and *Parintins* Cultural Complex	08/11/2018	*Amazonas*	estadual
14. *Bembé* of the market	13/06/2019	*Bahia*	local—*Bahia*

In this aspect, the emptying of the countryside and the migration of the population to the city modified the very practice and representation of the foliage of kings, as the party gained a public space and an organizing team, which enabled the resignification of some festive elements, as in the case of the party character (Goulart 2018).

In a Brazilian *Folia de Reis*, the party is usually the one that fulfills a promise to the Holy Kings, offering the place, as well as all the organization and support of extra expenses of the party. Therefore, the group of singers, clowns, and other devotees and sympathizers of the celebration is responsible for the removal of the flag from the house of this party on December 25th, and the departure for the well-known *giro* of *Folias de Reis* (peregrination). On this day that lasts until January 4 or 5, they visit various houses, taking the blessings of the Saints, and in exchange, collect gifts (food, drinks, money) that will be converted into the great final banquet (meeting of flags), celebrated on January 6th—the day of *Santo Reis* (Saints Kings or Three Kings).

As measured, with the changes in the history of the place, there was also a change in the senses attributed to this central character of the feast. Currently, in *Florínea*, any individual/family who has a devotion to the Saints Kings can assume the role, because the responsibility of organizing the party—which today has a large audience—as well as the availability of the place, is for the team that composes the Association. Another important point is that the stabilization of the party in the city and the union of the two festive groups was only possible because of the negotiation between the community and the local city hall, which allowed: the purchase and reform of the celebration space; the liberation of the revelers who are public servants to perform the peregrination; and the supply of school buses that take the revelers to visit different cities and rural districts (Goulart 2018).

These attitudes, although not a public policy, have contributed to the strengthening and reconfiguration of the local identity of the *Folia de Reis*. In the words of Rovai (2016, p. 249),

> The function of public policies should not only aim to put culture within everyone's reach but to contribute so that communities can experience and value their own culture, choosing the ways to experience it, spread it, and preserve it in what makes sense and gives it life.

It is important to say that from this research, more groups of *Folia de Reis* of cities in the west of São Paulo state were mapped, and as observed, there is a growing organization of communities in the region, either in informal commissions or registered associations, in the sense of mobilizing popular forces to give continuity to cultural practice, which requires, as already announced, the constant articulation of the *foliões* with the members of the local public power.

This point is relevant because, among the records made by IPHAN, except for the Party of *Divino Espírito Santo* in the City of *Paraty/Rio de Janeiro* (6) (requested by the Historical and Artistic Institute of *Paraty*) and the Cultural Complex of the *Boi Bumbá* in the Middle Amazon and Parintins (13) (requested by the Secretary of Culture of the State of *Amazonas*), all the others were requested by civil Associations directly linked to cultural assets, which characterizes the possibilities of safeguarding these references, since they indicate the public interest in the demand for the identification, appreciation, and transmission of intangible heritage.

Not by chance, in the I and II Evaluation Meeting of Safeguard Plans and Actions as Cultural Heritage of Brazil, held in 2010 and 2012, the strong presence of the communities that already had their assets registered was observed, leading to discussions about their particular demands, which included the manifestation about the importance of the political articulation of the holders of cultural assets in the face of the precariousness of public policies in different areas, which may directly affect cultural heritage. In this aspect, the meetings were also clarifying the need for dialog between the powers of the Union, states, and municipalities in the constitution of public policies (IPHAN 2018, pp. 23–24).

Regarding the particular demands of the communities, the Dossier of the Party of *Divino Espírito Santo* of the *Pirenópolis* (2) presents an interesting caveat. In the document, a conflict of interests experienced in the festive community was reported, when the church expressed itself in favor of choosing the Emperor of the Divine based on the particular religiosity of the individual who would assume that important character of the celebration, and also when the municipal public power began to spectacularize the party, making it the object of mass tourism (IPHAN 2017, pp. 112–117).

These interventions, not always in line with the wishes of the local holders of the cultural good, indicated, on the other hand, the importance of the heritage policies that were to be constituted in the place. Thus, taking into consideration community demands, alternatives were presented regarding the safeguarding of the Feast of the Divine, which was not restricted to the official registration and recognition of this good. These alternatives were not restricted to the registration and official recognition of this good:

> I. encourage, through public and business initiatives in *Pirenópolis*, a permanent commitment to respect local culture and to guarantee the spaces of expression of the Festival of the Divine;
>
> II. create mechanisms for consultation with the population, so that it may express its opinion on the modes of protection and maintenance of the feast that it deems appropriate, guiding and strengthening the construction of protection policies and safeguard measures;
>
> III. regulate tourist activities in the municipality in order to promote cultural tourism and respect for the activities of the festival;
>
> IV. set up a music school or conservatory to strengthen the city's musical vocation and support local talents;
>
> V. demand from the competent authorities the implementation of public infrastructure, health, and education policies. (IPHAN 2017, p. 118).

From the excerpt, alternative V stands out, as it enables an unrestricted reflection on cultural heritage policies. Basically, it guides that public policies should be thought in conjunction with the other demands of society. There is no way to conceive, for example, the social awareness about heritage, and consequently, about the place of speech of the various social groups of a single locality, without the instrumentalization and application, in this case, of education policies based on ethical and democratic responsibility.

If there is not the constitution of a responsible society, with public policies directed to the valorization of the producers of cultural goods, and not only of products (Canclini 2015, p. 211), very probably the patrimonies will be at the mercy of time or subjugated to market interests, as signaled the Dossier of the Party of *Divino* in *Pirenópolis*.

It is important to stress this initiative of government institutions together with the communities in the preliminary analysis of the cultural asset because it was in this process of dialog that the registration of the state complexes that have the

Boi (Bull) as their main element also emerged. In the same way, it occurred with the enlargement of the festive territory itself in the registration of the Glorious *São Sebastião* festivities in the *Marajó* region (8), previously only the city of *Cachoeira do Arari* was delimited.

It was reported in the Opinion of the Party of St. Sebastian (62/2013/CGIR/DIP/Iphan), that in a Seminar held in 2007, in the city of *Cachoeira do Arari*, it was identified the fragility in which the cultural asset was found in some cities of the region of *Marajó*, which could hinder the perpetuation of the folias and knowledge from it. Thus, it was decided to make a broader record of intangible heritage in order to give continuity to the axes of social and political mobilization and participatory management, which could contribute to a greater sustaining of the local identity and sustainability of the festival.

Another important point is that, although the registration of the ox complexes in *Maranhão* and *Amazonas*, territorially extensive and culturally multiple states, the specificities (*Sotaques*) of some groups were signaled, which praised the richness of the distinct elements and manifestations present in these celebrations that, at the same time, dialog with the beliefs and rhythms established in the Christian, Afro-Brazilian, and indigenous religions.

In turn, in Opinion 07/12/CR/CGIR/DPI/Iphan, referring to the Registration Process of the Party of the Divine Holy Spirit of *Paraty*, the following problem was presented: "[...] how to approach those feasts whose devotion is recurrent throughout the country, and how to evaluate and select those that can be recognized as intangible heritage of Brazil" (p. 5). A subject where no precise conclusion was reached, but which in the document itself, as well as in the other Opinions and Dossiers of the immaterial patrimonies registered according to Table 1, signaled a relevant justification: the specific historical characteristic of each city and communities to which they belong. The Feast of the Divine of Paraty appeared, for example, in the eighteenth century. Already the *Pirenópolis*, in the nineteenth century.

From this long duration that marks the time of the foundation of the celebrations in the communities, comes the evidence that both *Paraty* and *Pirenópolis*, were taken as a material patrimony of Brazil already in the twentieth century. These observations, however, demonstrate the tendency of the public agency in the process of ascertaining immaterial goods to be inventoried and recognized at the federal level.

In the study by Abreu and Magno (2017, p. 21), there is a demonstration that one of the first initiatives of the DIP/IPHAN/RJ, regarding the Inventory of the *Folias de Reis Fluminenses*, was the selection of 15 of the 92 cities that owned the cultural good in the state. In this aspect, the initial criterion used for the selection of the *Folias de Reis Fluminenses* was the integration of the 15 cities with the first Growth Acceleration Program—PAC Historic Cities.

Like *Paraty* (Rio de Janeiro), the cities of *Pirenópolis* (*Goiás*), *Salvador* (*Bahia*), and *Belém* (*Pará*) were also awarded by PAC Historic Cities, which not only confirms what was suggested above, but also leads one to believe the importance of linking this Program, at the stage of preliminary analysis, to the possible intangible heritage of the places. In Table 1, we have already identified, for example, that the Party of the Divine Holy Spirit of *Pirenópolis*/*Goiás* (2) was registered on 05/13/2010, the

Party of *Senhor Bom Jesus* do *Bonfim/Bahia* (7) on 06/05/2013, the Glorious Feast of São Sebastião in the region of *Marajó/Pará* (8) on 11/27/2013, the Car Parade of Ox of the *Party* of the Divine Eternal Father of *Trindade/Goiás* (11) on 09/15/2016 and the *Bembé* do Market/*Bahia* (14) on 06/13/2019.

On this characteristic of long duration present, sometimes, in the intrinsic relations between material and immaterial goods, there are the problems of establishing a more precise means of patrimony, for example, the *Folia de Reis* do West of *São Paulo*. There is a deadlock between a celebration that has a core of devotion and cultural practices that recur in much of the national territory, and at the same time, has a specific identity in this region of *São Paulo*, populated mainly since the beginning of the twentieth century, and lacking, even, material heritage policies (Nascimento 2012).

On this limit, ascertained from the *Folia de Reis* of *São Paulo*, attention is paid to the importance of the movement for the constitution of community associations, already ascertained in the region, as well as the importance of studies and academic projects that focus not only on the registration of the presence of popular celebrations in these places, but also contribute to the demands of society, with proposals and reflections on public policies in a sense already mentioned by Rovai (2016, p. 249).

Through his experience with the *Jongo*, Calabre (2016, p. 263) demonstrates, in the meantime, that although the instrumentalization of Brazilian intangible heritage policies has had a positive impact on the society that holds the cultural good, it is still necessary to reflect on possible scenarios of financial and political crisis, which signals the relevance of the independence of groups and communities before the powers of the State.

4 Conclusion

From the material to the immaterial, there has been a long historical trajectory on the concept of heritage, a trajectory that has positively led to the promotion of public policies that enable the recognition, safeguarding, and transmission of cultural goods from generation to generation.

In the case of celebrations, through the study of documents referring to their records as intangible heritage of Brazil, we can identify them, fundamentally, as cultural goods of a religious nature. In this aspect, we evaluate them within their limits of patrimonialization, since in a single festival, there may be beliefs and values attributed in different ways, as is the case of the Ox Complexes in *Maranhão* and *Amazonas*, or even the *Folia de Reis Fluminenses*. Moreover, within these limits, or even possibilities, we identify its patrimonialization linked to the time of existence of the cultural good, which frames it to a remote and traditional past, in a sense stipulated by Hobsbawm and Ranger (2018, p. 9), of invariability. This means that there is still a paradox to be faced in the politics of intangible heritage since the legislation itself regarding cultural goods suggests its dynamism.

Regarding the possible impacts of the IPHAN methodology on the *Círio de Nazaré*, the only cultural good that has already passed the 10-year period—time stipulated for the application and analysis of the federal agency's safeguarding policies—unfortunately, we have not been able to obtain opinions and reports that should be available on the Institute's Web Site.

In any case, the other documents pointed out in the text, as well as the presentation of studies and impressions on the various cultural assets that are part of the history of communities, led to the understanding of the importance of the articulation of the groups themselves in relation to the preservation of their cultural assets—the possibility of safeguarding already indicated through the new Civil Associations. Not by chance, the proposals for the registration of most of the celebrations registered as an immaterial heritage of Brazil, emerged from these local organizations.

References

Abreu R, Magno M (2017) Desafios na patrimonialização de bens imateriais de caráter religioso: o caso das Folias de Reis Fluminenses. Religião e Sociedade, Rio de Janeiro, vol 37, no 3, pp 18–45, Dec 2017. https://doi.org/10.1590/0100-85872017v37n3cap01. Accessed 25 Feb 2020

Brasil [Constituição (1988)] Constituição da República Federativa do Brasil: texto constitucional promulgado em 5 de outubro de 1988, com as alterações adotadas pelas Emendas Constitucionais nos 1/1992 a 68/2011, pelo Decreto Legislativo n° 186/2008 e pelas Emendas Constitucionais de Revisão nos 1 a 6/1994, vol 35. ed. Câmara dos Deputados, Edições Câmara, Brasília

Calabre L (2016) O lugar da cultura popular nas políticas públicas: ações no campo do patrimônio imaterial. In: Mauad AM, Almeida JR, Santhiago R (eds) História pública no Brasil: sentidos e itinerários. Letra e Voz, São Paulo

Canclini NG (2015) Culturas híbridas: estratégias para entrar e sair da modernidade. Edusp, São Paulo, p 2015

Cunha FL (2004) Da Marginalidade ao Estrelato: o samba na construção da nacionalidade (1917-1945). Editora Anna Blume, São Paulo

Cunha FL (2009) As matrizes do samba carioca e carnaval: algumas reflexões sobre patrimônio imaterial. Patrimônio e Memória (UNESP), vol 5, pp 1–23

Fonseca MCL (2017) O Patrimônio em processo: trajetória da política federal de preservação no Brasil. Editora UFRJ, Rio de Janeiro

Funari PP, Pelegrini SCA (2009) Patrimônio Histórico e Cultural. Jorge Zahar Editora, Rio de Janeiro

Goulart RS (2018) Sentidos da Folia de Reis: um estudo da memória e da identidade da celebração popular em Florínea/SP. Alameda, São Paulo

Goulart RS, Cunha FL (2018) As folias de reis em Ourinhos e Salto Grande (SP): por que e como estudar? Geografia e pesquisa (UNESP. OURINHOS), vol 12, pp 42–53

Hobsbawn E, Ranger T (2018) A invenção das tradições. Paz e Terra, Rio de Janeiro

IEPHA (2016) Dossiê para Registro das Folias de Minas. Belo Horizonte, IEPHA

IPHAN (1937) Decreto-Lei N° 25, de 30 de novembro de 1937. IPHAN, Rio de Janeiro

IPHAN (2000) Decreto-Lei N° 3.551, de 4 de agosto de 2000. IPHAN, Brasília

IPHAN (2006) Decreto N° 5.753, de 12 de abril de 2006. IPHAN, Brasília

IPHAN (1997) Carta de Fortaleza, 14 de novembro de 1997. IPHAN, Brasília

IPHAN (2017) Festa do Divino Espírito Santo de Pirenópolis. Série Dossiês Iphan, vol 17. IPHAN, Brasília

IPHAN (2000b) Inventário nacional de referências culturais: manual de aplicação. IPHAN, Brasília

IPHAN (2015) Portaria N° 299, de 17 de julho de 2015. IPHAN, Brasília
IPHAN (2016) Portaria N° 200, de 18 de maio de 2016. IPHAN, Brasília
IPHAN (2018) Saberes, fazeres, gingas e celebrações: ações para a salvaguarda de bens registrados
    como patrimônio cultural do Brasil 2002-2018/Instituto do Patrimônio Histórico e Artístico
    Nacional (Brasil); coordenação de edição Rívia Ryker Bandeira de Alencar. IPHAN, Brasília
Meneses UTB (2012) O campo do patrimônio cultural: uma revisão de premissas. In: IPHAN.
    I Fórum Nacional do Patrimônio Cultural: Sistema Nacional de Patrimônio Cultural: desafios,
    estratégias e experiências para uma nova gestão, vol 1, Ouro Preto (MG). IPHAN, Brasília
Nascimento RM (2012) A preservação do patrimônio cultural no oeste paulista. Paço Editorial,
    Jundiaí
Pollak M (1989) Memória, Esquecimento, Silêncio. Estudos Históricos, Rio de Janeiro 2(3):3–15
Rovai MGO (2016) Políticas culturais e tradição popular: uma reflexão sobre os caminhos trilhados e
    sonhados. In: Mauad, Ana Maria; Almeida, Juliene Rabêlo de, Santhiago, Ricardo, 2016. História
    pública no Brasil: sentidos e itinerários. Letra e Voz, São Paulo
UNESCO (2003) Convenção para a Salvaguarda do Patrimônio Cultural Imaterial. Paris, 17 de
    outubro de

Cultural Landscape and Tourism

Perception of the Cultural Landscape in Historical Centers

Rosio Fernández Baca Salcedo

Abstract The cultural landscape impacts our senses, our activities, interferes in decision-making, judgment, and values and in communication with the urban space. The interrelated dimensions, necessary for understanding the perception of the cultural landscape are the physical, social, and symbolic dimensions. In this article, we address the perception of the landscape in historical centers in the physical (categories: group/dispersal, integrate/segregate, attract/repel, open/close, walk, sit, and stand) and social (categories: necessary, optional, and social) dimensions.

Keywords Cultural Landscape · Historic Center · Perception

1 Introduction

The configuration of the cultural landscape in historical centers impacts our senses and interferes in decision-making, values, and our relationship with the urban space.
Historic centers express the initial layout of the city:

> They are urban and architectural structures that express the political, economic, social, cultural, and technological manifestations of the social formations of the different historical periods, through which it has evolved, unitary or fragmentary structures, although they have been transformed over time and present themselves as testimonies of civilizations of the past (Salcedo 2007, p. 15).

Furthermore, historical centers have a "special value as historical testimony or particular urban or architectural features" (Government of Italy 1972 apud Cury 2004) (Fig. 1).

The concept of cultural landscape comes from cultural geography and refers to the results of the interaction between human actions and the primary landscape that develops in time. Landscape can be understood as "the formal expression of the

R. F. B. Salcedo (✉)
Departament of Architecture, Paulista State University "Julio de Mesquita Filho"– UNESP, São Paulo, Brazil
e-mail: rosio.fb.salcedo@unesp.br

© The Author(s), under exclusive license to Springer Nature Switzerland AG 2021   101
F. Lopes da Cunha and J. Rabassa (eds.), *Festivals and Heritage in Latin America*,
The Latin American Studies Book Series,
https://doi.org/10.1007/978-3-030-67985-9_7

Historical Center of Cuzco (Peru), recognized as World Heritage by UNESCO in 1983.

Historical Center of Ouro Preto (Brazil), recognized as World Heritage by UNESCO in 1980.

Fig. 1 Historical Centers

numerous relationships existing in a given period between an individual or society and a topographically defined territory, whose appearance is the result of special action or care, natural and human factors, and a combination of both" (Council of Europe 1995 apud Cury 2004, p. 331). Rapoport (2003, p. 53) emphasized that the cultural landscape refers to the results of the interaction between human actions and the primary landscape that develops in time. The more the landscape is modified by men, the more cultural it is.

Still on the cultural landscape in historical contexts, in 2009, on a national scale, Ordinance No. 127 of April 30, 2009, established the seal of the Cultural Landscape and in Article 1 of Title 1, it defined:

> Brazilian Cultural Landscape is a peculiar portion of the national territory, representative of the process of interaction between man and the natural environment, to which life and human science have imprinted marks or attributed values. Single paragraph - The Brazilian Cultural Landscape is declared by a seal instituted by IPHAN, through a specific procedure.

Therefore, the valorization and protection of the Brazilian cultural landscape are declared by seal by the Institute of National Historical and Artistic Heritage (IPHAN). In Brazil, the city of Rio de Janeiro was recognized as Cultural Landscape by UNESCO on July 1, 2012; and Canudos in Bahia was recognized as a Brazilian Cultural Landscape by IPHAN (Fig. 2).

The Cultural Landscape is not static. Because it is dynamic, it suffers the consequences of a chronotopic order in unstable equilibrium, ceaselessly submitted to the deepest forces of survival of our species (Muntañola, no date available). Thus, the cultural landscape can be read, understood, intervened through the chronotopes sedimented in time and space. Every intervention alters the configuration of the landscape, and consequently, its perception. Like Paulista Avenue in the city of São Paulo, inaugurated on December 8, 1891, at the beginning of the twentieth century it was characterized by earthly constructions and townhouses and on the road streetcars and vehicles circulated. Decades later, these constructions were replaced by buildings and the road gave way to the traffic of vehicles and buses, changing the landscape of the beginning of the twentieth century (Fig. 3).

Rio de Janeiro (Brazil), Cultural Landscape declared by UNESCO, on July 1, 2012.   Canudos in Bahia, Brazilian Cultural Landscape proposed by IPHAN.

Fig. 2 Cultural Landscapes in Brazil

Paulista Avenue (São Paulo, Brazil), between 1896 and 1900. Guilherme Gaensly/Reproduction Acervo Light.   2019 Paulista Avenue in São Paulo (Brazil)

Fig. 3 Paulista Avenue in the city of São Paulo (Brazil)

Therefore, the primary landscape is continuously built and transformed by culture. Culture is a complex unit that comprises knowledge, faith, art, laws, customs, and habits acquired by humans as members of society (Rapoport 2003, p. 131). Culture is a dialog, exchange of ideas and traditions, transmitted from generation to generation, which, in isolation, runs out and dies. "Each culture represents a unique and irreplaceable set of values, since the traditions and forms of expression of each people constitute their most finished way of being present in the world" (ICOMOS 1985 apud Cury 2004, p. 272).

The cultural landscape in historical centers seduces the diverse historical, cultural, social, economic, political, and technological manifestations of the social formations of the different historical periods through which it evolved. Therefore, its exceptional value from the point of view of history, art or science, must be safeguarded to preserve history, identity, and memory.

Cultural identity is understood as the "richness that dynamizes the possibilities of realization of the human species by mobilizing each people and each group to nourish themselves from their past and to collect the external contributions compatible with their specificity and thus continue the process of their own creation" (ICOMOS 1985 apud Cury 2004, p. 272).

Collective memory, on the other hand, can be understood as an "organized system of memories whose supports are the socio-spatial and temporal groups" (Menezes 1992, p. 15). However, national memory "is of the order of ideology. It is the soup of culture par excellence for the formulation and development of national identity, ideologies, national culture, and therefore, for the historical knowledge of these phenomena" (Menezes 1992, p. 15).

The configuration of the cultural landscape interferes with our perception and cognition. Oliveira and Machado (1989) emphasized that:

> In the interactions between man and landscape there is a continuous exchange and mutual influence between the exterior world and the interior personal world. Thus, the inner and outer worlds are always interconnected in the functioning of a human organism; they interact and evolve together, and the functional exchanges between the individual and the outer environment include two interdependent aspects: the cognitive and the affective. Cognitive life and affective life are inherent, although distinct. They cannot be separated because every exchange with the exterior supposes at the same time a structuring and an appreciation, and one cannot be reduced to the other (Oliveira and Machado 1989, p. 43).

Perception will always be linked to a sensory field and will consequently be subordinated to the presence of the object, which offers it knowledge by immediate connotation" (Del Rio and Oliveira 1996, p. 203). Perception deals with the relationship between the environment and its users and the stimuli provoked by this environment on the senses of its users: sight, hearing, smell, touch, and gustation (Salcedo 2016).

Tuan (1983) pointed out that the perception of the built space is extremely varied:

> But the ways in which people perceive and evaluate this surface are more varied. Two people do not see the same reality. Neither two social groups make exactly the same evaluation of the environment (…). All human beings share common perceptions, a common world, by virtue of having similar organs (Tuan 1983, p. 151).

Therefore, how we relate to the landscape will depend on our culture, our interests, and our needs. However, cognition is: "the mental process by which, from interest and need, we structure and organize our interface with reality and the world, selecting perceived information, storing it, and giving it meaning" (Del Rio and Oliveira 1996, p. 203).

> The meaning that is attributed to the perceived object can be different from person to person, and is in function of individual characteristics, values, symbols, customs, culture, personality, temperament, age, sex, income, social classes and origin, among others (Salcedo 2016, p. 71).

According to Rapoport (2003, p. 27), the cognitive can be understood as the mental process that intervenes between perception (acquisition of information) and knowledge about the surroundings. Anthropology understands cognition, relative to the construction of the world, as its categorization in different domains and nomination (rationalization of the world by the human being). In the most common psychological meaning, it describes how we learn things about the world, build mental maps, orient ourselves, and navigate, that is, how we operate in the world. At the same time, the affective question includes the emotions, moods, sensory qualities, meanings, among others, provoked by the environment (Rapoport 2003, p. 28).

It is interesting to know how and who shapes the cultural landscape. Article 30 of Chapter IV of the 1988 Constitution of the Federative Republic of Brazil, stated that it is the Municipality's responsibility to "promote, where appropriate, territorial planning, through planning and control of the use, parceling and occupation of urban land". Still, when it comes to cultural landscapes in historical centers, it is the heritage preservation institutions that intervene in them, preserving or not the architectural and urban heritage. Also, in the construction of the landscape, the owners of the means of production intervene, especially the great industrialists, the landowners, the real estate promoters, and the social groups or social movements. The action of the agents in the production of the landscape is framed within a legal framework that regulates their action. The legal framework reflects the dominant interest of one of the agents, allows the reproduction of productive relations, and the continuity of the accumulative process.

The legislation, urban norms, and master plan in each city are defined by the Municipal Government. Land use, collective equipment, basic sanitation, road infrastructure, among others, determine the cost of land and construction. Thus, the configuration of the landscape is regulated and managed by public management, through urban norms, the master plan; policies, programs, and projects. It is up to the Municipal Government through the Master Plan, the offer of collective equipment and services, infrastructure, among others, to improve the quality of urban life and social welfare for all, to preserve the cultural heritage and natural resources, in order to configure living, safe, sustainable, and healthy cultural landscapes.

2 Dimensions of the Cultural Landscape

The interrelated dimensions, necessary for understanding the perception of the cultural landscape and the place are the physical, social, and symbolic dimensions. In this article, the physical and social dimensions are addressed.

2.1 Physical Dimension of the Cultural Landscape

The quality of the physical dimension of the cultural landscape impacts our senses and interferes in our perception, decision-making, judgment, values, conduct, and communication, with continuous interaction between humans and the landscape.

The physical dimension of the cultural landscape is composed of the elements of nature, fixed and semifixed (Fig. 4).

The elements of nature are: water (rivers, rain, sea, etc.), land (soil), fire (sun, heat, flame, etc.), air (wind, whirlwind, etc.), flora and fauna (Oliveira and Machado 1989).

The fixed elements are composed of buildings, walls, floors, roofs, electricity, water, and sewage networks, among others. The semifixed elements include urban furniture, visual communication, and vegetation (Rapoport 2013, p. 44).

The spatial organization of buildings and public spaces of the physical dimension of the cultural landscape can result in the following categories: grouping/dispersing, integrating/secreting, attracting/repelling, opening/closing, walking, sitting, and standing.

Grouping/dispersing: it is necessary to group people for social activities. For this, it is essential to design buildings with harmonious façades, rich compositions, especially on the first floors, which is the height at which our vision reaches and how far we can see and experience. Also, it is important to be able to walk from one point to another (Gehl 2013, pp. 94–95).

Integrate/secure: integration implies that various activities and categories of people can work together, side by side. Therefore, people who work and live in different buildings could use the same public spaces and meet each other when developing everyday activities. Segregation implies a separation of urban functions and groups that differ from each other (Gehl 2013, p. 113).

Attract/remove: public spaces and residential areas of the city can be attractive and easily accessible, and encourage people and activities to move from the private

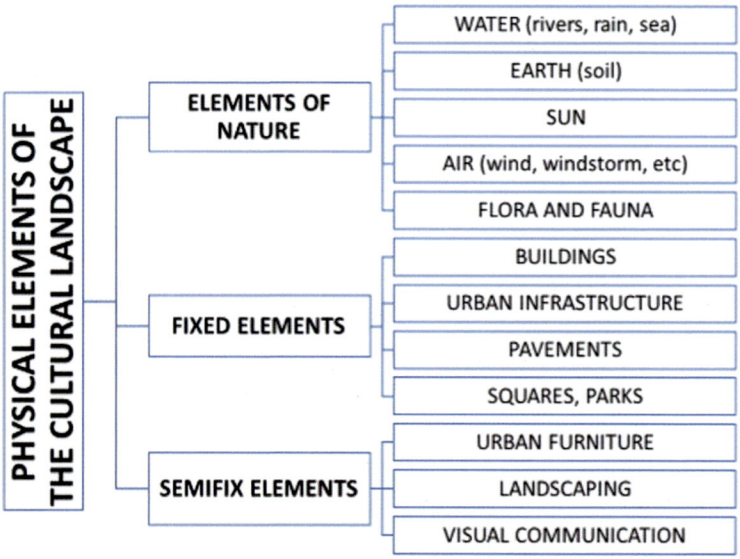

Fig. 4 Physical elements of the cultural landscape. *Source* elaboration by the author

to the public environment. On the other hand, public spaces can be designed in a way that makes it difficult to enter them, physically and psychologically. Attracting or repelling depends on how the public space is located in relation to the private and how the zone bordering them is designed (Gehl 2013, p. 125).

Open/close: the contact through experience between what is going on in the public surroundings and what is going on inside commercial, service, institutional, and financial buildings, can increase the enrichment of the possibilities of experiences, in both directions. Opening to have an exchange of experiences in two directions is not only a question of glass and window, but also of distance (Gehl 2013, p. 133). In this sense, creating buildings with few squares facing the street or blind walls avoids experiences between people in buildings and people in public spaces.

2.2   Social Dimension of the Cultural Landscape

The activities performed in the public spaces of the streets, avenues, squares, gardens, and parks, are conditioned by the quality of the space. According to Gehl (2013, p. 17), outdoor activities in public spaces can be divided into three categories: necessary, optional, and social.

Necessary activities are all those in which people are more or less obliged to do, such as going to school and work, waiting for the bus or a person, going shopping, among others. In these activities, people have no choice, they happen frequently throughout the year, regardless of the quality of the public space.

The optional activities take place in public spaces when the external conditions are favorable, when the weather and the place invite you to do it, like: sit, contemplate, eat, play, walk on a sidewalk, stop, sit to enjoy the view or the good weather, stand and have a good look at the city, etc. (Gehl 2013, pp. 17–19). When the public space is of poor quality, optional activities take place only when necessary. When the public space has quality, the optional activities happen with greater frequency and duration.

Social activities require the presence of other people and include all forms of communication between people in the public space (Gehl 2013, p. 20). These activities take place in the public or semi-public spaces of the homes, work, school, leisure. They vary according to the quality of the urban context and can be children's games, conversations, contacts, community activities, dating, street parties, meetings, parades, and political manifestations. These activities take place in private spaces outside buildings such as gardens, terraces, workplaces, among others. The analysis of this type of activity takes place in the outdoor spaces accessible to the public.

3 Perception of the Cultural Landscape

The genesis of cities created until the beginning of the twentieth century configured landscapes in scales at eye level. The historical centers of origin of the city sheltered the commerce, services, finances, and residences, the constructions were compact, the circulation between the place of work or commerce or services and the residence was made on foot, from there that the cities were adapted to the senses and the potential of the human beings. In today's urban planning, the traditional knowledge of scale and proportions of buildings was gradually lost, resulting in newly built urban areas, often on a scale far removed from what is perceived as significant and comfortable (Gehl 2015, p. 55).

The human body, its senses, and its mobility are the key to good urban planning for all. The challenge is to build splendid cities at eye level, with large buildings rising above beautiful lower floors (Gehl 2015, p. 59).

The transition spaces in public places and the active or passive façades of the buildings around them interfere in the performance of activities. There is a direct connection between smooth transitions and living cities. The renovation of buildings to adapt them to commerce or services has led to the removal of the original doors from first floor facades to replace them with metal rolling doors, giving way to passive façades. Meanwhile, the preservation of historical building façades configures harmonious and active landscapes, providing a greater flow of people in transition spaces.

Also, the enormous distance between tall or large buildings marks an impersonal, formal, and cold urban environment, which only serves to carry out mandatory activities. At the same time, low-floor buildings with shorter distances create more cozy environments, providing optional and social activities.

By walking along the sidewalk, through our vision we give interest and intensity only to the façades of the first floors of buildings. If the façades of the first floors are rich in variations and details, our urban walks will be equally rich in experiences (Gehl 2015, p. 41). Therefore, ground-floor, house, and low-floor buildings have an impact on perception because the vision allows us to see them. However, tall buildings will be more difficult to experience, and can only be appreciated from a distance and never up close.

The "life between buildings", in cities, ensembles, centers and historic neighborhoods, with dense traffic and narrow sidewalks, it makes it impossible to carry out optional and social activities. As living, safe, sustainable and healthy cities, the prerequisite for the existence of urban life, offer good opportunities for walking. However, the broader perspective is that a multitude of valuable social and recreational opportunities arise when life on foot is strengthened (Gehl 2015, p. 19). Also, the incidence of quality improvement in everyday and social activities in cities can be seen where there are pedestrian streets or traffic-free zones, children play on the sidewalks and on the street, and at the entrances to houses and balconies are widely used to be abroad and to make a wide network of contacts among the neighborhood.

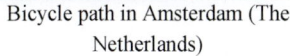

Bicycle path in Amsterdam (The   Pedestrian street in Curitiba (Brazil)
Netherlands)

Fig. 5 Public spaces for hiking and bicycle paths

In public spaces with physical infrastructure in the form of quality walking trails, bicycle paths, areas for cultural and recreational activities, they enable the performance of various activities: necessary, optional, and social (Fig. 5).

The concentration of commerce, services, and finances to the detriment of the residential function creates underutilized public spaces, where the streets are only used during the hours of these equipments and after working hours the streets become empty and unsafe. However, the increase of housing units in groups, centers, and historical neighborhoods increases the number of inhabitants. The 24-hour use in these areas provides various activities such as restaurants, pharmacies, bakeries, schools, gyms, among others. The look of people in the street creates security.

The quality of urban rivers also affects our perception and the activities we carry out around them. The river channels human waste, industrial waste, among others, and when not treated becomes a dead river, besides contaminating the environment. However, the open river, treated and with recreational spaces along its course, becomes an area for the realization of optional and social activities densely occupied. We can highlight the Seine River in Paris, which "every year in summer, the avenue along the Seine River in Paris is closed and converted into 'Paris Beach', quickly invaded by thousands of inhabitants" (Gehl 2015, p. 17) (Fig. 6).

The microclimate and the quality of public space interfere in decision-making for activities. "Climate and comfort vary with the seasons and geographical location. The sun is a great attraction in temperate regions, while shade is an appreciated quality in warmer climates" (Gehl 2015, p. 169).

Grouping/dispersing: a street with continuous buildings and pedestrian-oriented entrances represents a clear and congruent grouping. On the contrary, a street in a condominium with a blind wall disperses people (Fig. 7).

Attract or repel depends on how the public space is located in relation to the buildings and how the boundary zone between them is drawn (Gehl 2013, p. 125). Example: Houses with permeable outdoor gardens, transition areas between public and private spaces attract optional and social activities. However, buildings aligned

Siena River in Paris (France)   Tietê River in São Paulo (Brazil)

Fig. 6 Urban rivers

Historical center of Paraty (Brazil)   Condominium wall in Bauru, São Paulo
                                                                (Brazil)
                                                     Source: Salcedo (2020)

Fig. 7 Group/Dispersal

with sidewalks or walls facing the street, with no private–public transition areas, repel the realization of optional or social activities (Fig. 8).

Open/Close: to have an exchange of people's experiences in two ways, both in public and private spaces (buildings). The buildings should be around the public space and the façades of these buildings facing the street should be permeable. For example, in commerce, services with glass windows allow the visualization of activities inside (Gehl 2013, p. 133) (Fig. 9).

Walking is more than just moving around for necessary optional or social activities, going to work, school, shopping or just walking. Walking requires quality landscapes, quiet, safe, good views, nonslip floors, urban furniture, vegetation. "One can choose between walking on a deserted street or a lively street, most will choose a lively street" (Gehl 2013, p. 31).

The cultural landscapes with quality: quiet, safe, urban furniture, good views, and protected from wind, sun, and rain; provide places to sit, chat, wait, contemplate, among others. "Choosing where to sit also depends on the bench that offers a good

Neighborhood in London (England)   Buildings with walls to the street, Bauru, São
Paulo (Brazil)
Source: Salcedo (2020)

Fig. 8 Attract or Repel

view, they are used more than with benches with little sight". People prefer places
where there are people.

4   Concluding Remarks

A human cultural landscape, sustainable, alive, and safe is healthy, improves the
quality of urban life, impacts on perception and decision making for mandatory,
optional or social activities. If the configuration of the cultural landscape impacts
our senses, our activities, interferes with decision-making, communication with the
urban space, judgment, and values.

The safeguarding of the cultural landscape and the quality of urban life is
the responsibility of the public management, which should consider sustainable
economic development as a function of social development, create urban norms
and elaborate the master plan for adequate territorial planning, through planning and
control of the use, parceling and occupation of urban land, protection of the local
historical-cultural heritage, and preservation of natural resources. As well as the
elaboration of policies, programs, and projects of urban, architectural and landscape
intervention that meet the needs, expectations, and well-being of the population.

References

Brazil Portaria (2009) N° 127, April 30, 2009. It established the seal of the Brazilian Cultural
Landscape
Brazil (2001) Statute of the City, Law N° 10,257, of July 10, 2001
Brazil (1988) Constitution of the Federative Republic of Brazil, 1988

Council of Europe (1995) Committee of Ministers. Recommendation N°R (95)

Cury I (2004) On the integrated conservation of areas of cultural landscapes as part of landscape policies. In: Heritage letters, 3rd edn. IPHAN, Rio de Janeiro, pp 329–345

Gehl J (2015) Cities for people (transl by Di Marco A), 3rd edn. Perspectiva, São Paulo

Gehl J (2013) La humanización del espacio urbano: la vida social entre los edificios. Reimp, Barcelona, p 2013

Government of Italy (1972) Letter of Restoration, 1972. In: Cury,, I., 2004. Letters of Heritage. 3rd edn. Ver. Aum. Rio de Janeiro: IPHAN, 2004. pp 147–170

ICOMOS (1985) Declaration of Mexico. World conference on cultural policies. In: Cury I (ed) Heritage letters, 3rd edn. IPHAN, Rio de Janeiro, pp 271–280

Menezes UTB (1992) The history, captive of memory. For a mapping of the memory in the field of Social Sciences. In: Revista Inst. Esta. Bras., São Paulo, 34-9-24

Muntañola J. The cultural landscape as dialogical landscape: an architecture for the future. In: Revista Lusofona de Arquitetura e Educação. No date available

Oliveira L, Machado L (1989) The perception of landscape as a methodology of geographical investigation. Anais do II Encuentro de Geografos de América Latina, Montevideo, pp 43–51

Rapoport A (2003) Culture, architecture and design. In: Arquitectonics magazine: mind, land & society, no 5. UPC, Barcelona

Rio V, Oliveira L (1996) Environmental perception. Studio Nobel and Editora UFSCar, São Carlos, São Paulo

Salcedo RFB (2007) Rehabilitation of residences in the Historic Centers of Latin America. Cuzco (Peru) and Ouro Preto (Brazil). UNESP Publishing House, São Paulo

Salcedo RFB (2016) Architecture and perception. In: Hipótese magazine, Itapetininga, vol 2, no 4, pp 70–89

Tuan YF (1983) Space and place. Difel, São Paulo

Tuan YF (1980) Topofilia. Difusão Editorial S. A, São Paulo

Landscape, Heritage and Tourism: Study in the Historic Center of Seville—Spain

Luciene Cristina Risso⊙

Abstract This chapter approaches the thematic of heritage landscapes and growing world tourism practices, taking as a case study, the historic center of Seville, with special emphasis on Triana. The main objective was to identify the main cultural heritage landscapes of the historical center of Seville and the pressures arising from the touristic industry. For this, the methodology involved readings, data collection, and geographical interpretations. In the survey of the primary data on the cultural heritage of Triana, there was an ethnographic description and participative observation of the Triana festivities (Holy Week, Santa Ana festivity, Flamenco show), and for the perceptual dimension, the leaders of the Northern Triana Neighborhood Association were interviewed. In the collection of the secondary data, the identification of the cultural heritage or assets of cultural interest ("Bien de Interés Cultural" – BIC) was used through the general catalog of the Andalusian historic heritage of the Council of Culture of the Junta of Andalucía (until 2016) and the tourism data were collected in the tourism council of Seville, referring to the years of 2017 and 2018. As a result, it is noted that most parts of the cultural heritage of the historic center fit in the typology of monuments legitimized by elites from different historical periods, which receive a massive amount of tourists, and in Triana, the ethnological heritages were a landscape potential, added to its history and symbolism, equally threatened.

Keywords History · Triana · Ethnological heritage · Residents · Monuments

L. C. Risso (✉)
Paulista State University "Julio de Mesquita Filho", Sao Paulo, Brazil
e-mail: luciene.risso@unesp.br

© The Author(s), under exclusive license to Springer Nature Switzerland AG 2021
F. Lopes da Cunha and J. Rabassa (eds.), *Festivals and Heritage in Latin America*,
The Latin American Studies Book Series,
https://doi.org/10.1007/978-3-030-67985-9_8

1   Introduction

The term landscape is multiple, since it is present in several disciplines; it is poly-semic, interdisciplinary, and contributes a lot to the heritage debate. As landscape and heritage are social/cultural heritance, there is a concern with the environmental status of these landscapes, integrant heritage, and people integrated into the land-scapes (at least it should be). It is from these experiences that values, belongings, meanings, identities, and individual/collective memories are constructed.

In this sense, studying landscapes means considering that it has several contexts, from geographical, geological, biological, artistic, touristic studies, and landscape interventions in Architecture to more philosophical researches considering the perceptual and existential sphere.

In general, in Geography, it can be said that landscape has materialities from nature and society, which allow us to interact and act on them, building new landscapes imbued with the juxtaposed immaterialities.

Using the framework of Cultural Geography in this research, we look at the symbolic landscape with the meanings linked to landscapes and places resulting from the relationship between society and Nature. Thus, we agree with Cosgrove (1998) and Nogué (2008) about the understanding of the landscape.

For Cosgrove, the landscapes are like texts to be decoded and for Nogué (2008, p. 19): "The landscape, from now on, will be conceived as a form, but also as a metaphor and as a system of signs and symbols".

To interpret a landscape, it is necessary to analyze how societies transform the natural forms, according to their symbolic culture and power. Here, power is consid-ered, since there is always a dominant culture, as the one that seeks to impose and reproduce its values on the landscape (Cosgrove 1998).

In Europe, the term landscape is widely used in the academic field and in politics and heritage management, especially after UNESCO adopted in 1992, the cultural landscape as an asset liable to be heritage, and conceptualized it as "combined works between man and nature" (UNESCO 1992). A category that aims to integrate the natural and the cultural.

In order to integrate the UNESCO World Heritage List, it is fundamental that the cultural landscape has exceptional universal (there are several critical authors of this issue, such as Veschambre (2008) and Di Giovine (2018), value and integrity/authenticity. The exceptional universal value, based on the registration criteria, must justify the worldwide importance that corresponds to the concepts of authenticity and integrity and to the forms of conservation and management.

Authenticity has also to do with the veracity of the cultural property and its attributes (form, design, materials, function, traditions, configuration, language, and spirit of the place), that is, when an asset expresses its true universal value. Integrity means that the asset is in good condition (integrity) and must be committed to preserving this state in face of changes over time. When this cultural landscape is recognized, after going through all the entire process at UNESCO, it is declared heritage.

Therefore, the choice to protect a heritage is not naive. It is intentional as stated by Veschambre (2008), as a mark of the past reactivated in the present (patrimonialization) is an intentional action by a certain social actor to legitimize, claim or challenge. When that form or mark of the past does not interest, it is erased (destroyed).

On the other hand, in the European Landscape Convention, the application of protection is not only directed at exceptional universal landscapes, as at UNESCO, but it is also concerned with the conservation and territorial management of the landscapes. For CEP (2000), the landscape is "any part of the territory as the population perceives, whose character is the result of the action and interaction of natural and/or human factors". The instrument is dedicated exclusively to European landscapes and, politically assumed, rekindled the debate and landscape practices. Therefore, the CEP concept of landscape is an advance, since it enriched the debate around a subjective plurality and of cultural conceptions.

In this way, European countries have an expressive heritage debate. In addition to following UNESCO guidelines for all the world patrimonialization process and management of the cultural assets, they have the CEP instrument to protect the European landscapes.

Spain joined, politically, to the CEP document in 2007. The Spanish law 16/1985, is the one that rules the historical heritage and is a reference for the other autonomous communities.

In the case of the Autonomous Community of Andalusia, the following types of heritage exist: (a) Monuments, (b) Historical Set, (c) Historical Gardens, (d) Historical Sites, (e) Archaeological areas, (f) Areas of Ethnological Interest, g) Areas of Industrial Interest, and h) Patrimonial Areas (Art. 25 of the Andalusian Law 14/2007).

Therefore, the references throughout the chapter on some of these heritage typologies will be based on the concepts of article 26 of that regulation. The research used, primarily, the collective typology of monuments, historical sets, and ethnological heritage.

Spain has a discussion and significant actions in the patrimonial area, since this country is in third place in the UNESCO list, with 48 registered assets, behind Italy and China (https://whc.unesco.org/en/list/stat). These heritages attract tourists from all over the world, which are increasing every year.

Tourism of world interest takes place in the post-war period, however, it is only from the 1970s that cultural tourism began, seen as an economic alternative for the countries.

According to the World Tourism Organization (UNWTO), in 2017, Spain won the second place among the most visited countries in the world with 81.8 million tourists (an increase of 8.9% compared to 2016), behind France, that received 86.9 million. The revenue collected was also in second place, with a turnover of around US$ 68 billion. British, German, and French people are among the foreign tourists that most visit Spain.

Landscapes and their heritage, as well as heritage landscapes, are increasingly attracting the tourism industry. It is a consumption that is impacting the places, the diverse heritages, and the lives of the residents due to the excessive flow of tourists.

The tourist activity has positive and negative impacts. Among the positive, that is, those that bring local benefits, are: the amount collected from the activity, the valorization and preservation of the historical heritage, urban cleanliness, accessibility, jobs, among others. Among the negative impacts are losses related to the environment, social exclusion, conflicts with the local community, etc. When the negative impacts begin to outweigh the positive ones, their effects may be irreversible.

In several European countries, it is noted the concern with this activity and the increased conflict between residents and tourists generating tourism phobia, due to the excessive number of tourists in the city and all the associated impacts, as in Amsterdam, Venice or Barcelona.

In Spanish tourism, the second most visited autonomous community in 2017, was Andalusia, behind Catalunya, with more than 22 million tourists (INE- Instituto Nacional de Estadística) In the autonomous community of Andalusia, tourism represents about 13% of the GDP.

In Seville, our study city with 688,711 inhabitants in 2018, received in that same year, 3,002,240 visitors (Consejo de Turismo de Sevilla 2018). It is a monumental city showing flamenco music and dancing, the Holy Week celebration, the Seville fair, and Spanish and regional gastronomy.

The historic center of Seville is one of the most extensive in Europe, with 335 hectares (3.35 km^2). It was classified as the historical set of Seville in 1990, also adding the extramural neighborhoods, such as Triana (Fig. 1). The typology of the historical set constitutes:

> the groupings of urban or rural constructions together with the geographical features that conform them as relevant for their historical, archaeological, paleontological, artistic, ethnological, industrial, scientific, social or technical interests, with sufficient coherence to constitute susceptible units of clear delimitation (Art. 26 of Andalusian Law 14/2007)

Tourists from several countries visit historical Triana, which is part of the historical set and it is highlighted by the history of ceramic art, navigation, and flamenco artists. Its urban landscape is challenging by itself due to the complexity of urban processes and dynamics that transform the places.

The main objective was to identify the cultural heritage of the landscapes of the historic center of Seville visited and the pressures arising from the tourist industry.

We tried to answer some questions such as: Which typologies predominate in the heritage assets in the historic center? Who legitimized them? Which heritage do the tourist visit most? In Triana, what do the residents think about their heritage and place?

The methodology involved readings, bibliographic review, primary and secondary survey. In the primary survey of the cultural heritage of Triana, there was an ethnological description and participative observation in the festivities (Holy Week, Santa Ana Festivity, and Flamenco shows) and for the perceptive dimension, a questionnaire was applied with the leaders of the North Triana Neighbors Association. In the secondary survey, the identification of cultural heritage or assets (BICs) was conducted using the general catalog of Andalusia historic heritage of the Culture

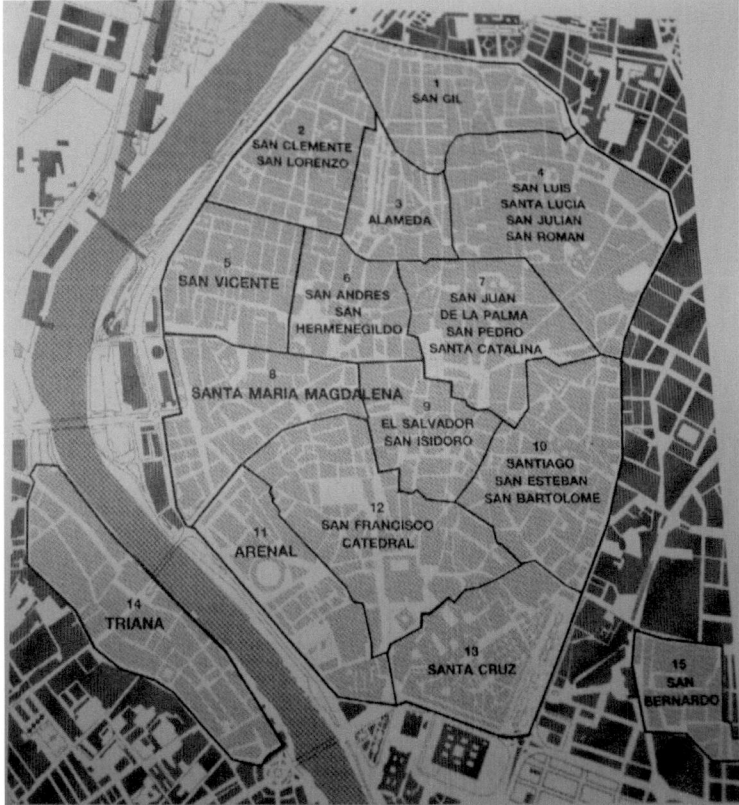

Fig. 1 Historical center of Seville (typology of Historical Set). *Source* Ortega (2006)

Council of the Junta de Andalusia (until 2016), and the tourism data were collected at the tourism council of Seville regarding the years of 2017 and 2018.

Finally, landscape, heritage, and tourism are intrinsically related, since UNESCO itself encourages tourism in the heritage landscapes. In the modern world, landscapes and their heritage have become objects to be consumed and sold in an inauthentic manner, but it is necessary that the local authorities discuss with their residents these issues in order to find new possibilities of ethical and sustainable interaction with landscapes and heritages.

2 Historical-Geographic Contextualization of the Historic Center of Seville

The physical-geographical base of Seville is characterized by the flood plains of the Guadalquivir River valley, delimited by mountain chains (Sierra Morena and the Sub-Baetic System).

The landscapes of the river and their fertile plains were essential for the construction of this territory. Several civilizations passed through this landscape. The foundation of this city is related to the Tartessian culture, which is called Seville as Hispal (XIIIth century BC) followed by the Phoenicians, Carthaginians, Romans, Visigoths, Arabs, and Christians. Romans called Seville as Hispalis and the river was called Betis (Center for Landscape and Territory Studies 2015). Arabians called it Ishbiliya (in Arabic: أشبيليّة) and then Shbiya, deriving to the current name. This Arab period from 711 AD was a period of Al-Andalus cultural wealth. In fact, around the year 713, a royal palace was constructed (the Real Alcázar), which the Almohade leaders used as a residence. It is precisely from this Moorish period that there are concrete records of Triana.

The history of the social formation of Triana is old. In Arabic *MaWara-Fnahr o Arrayana/Athriana* means "beyond the river". It was a suburb or neighborhood formed outside the walls of the main nucleus of Seville. In the Islamic era, the caliph Yusuf constructed the Bridge of the Barges in the year 1171, to better connect with the city of Seville, as well as serve as part of the defense of the city. Between 1220 and 1221, the Golden Tower was constructed in Seville with the same purpose.

From this dominant Almohade culture, there was a mosque that the kings used when they crossed the river by the Bridge of the Barges, constructed by them. They came to Triana for recreation, since they lived within the walls of Seville. During this period, there was also the installation of the handmade soap manufacturing of Queen Dona Juana (Ortega 2006). Due to the Christian invasion threat, the Almohade Empire built a fortress on the top of the cemetery (area of the Castle of San Jorge).

Since that time, Triana was an alternative landscape in which Cosgrove (1998) conceptualized, where the king's servants lived. The land was used for agricultural crops (cereals, grape, and olive), to supply Seville, as well as pottery activity, equally, corresponds to that time (Días Garrido 2004).

When the troops of Fernando III took Seville in 1248, the vital strategy was to stop the trade between Triana and Seville, leading to the fatal surrender of the Arabs.

After Christian domination, the Almohade fortress was reused, building on it the **Castle of San Jorge**. In 1266, the **Church of Santa Ana** began to be built in the Gothic Mudéjar style due to the amount of Christian people who populated Triana after the Reconquest. The castle was used in 1481, by the Spanish inquisition, which expelled the Jews from the city. It was the only church of the time that was built in this period, since the order was to reuse the mosques in Seville. The author gives an excellent example of Pople Gregory I "do not destroy the pagan temples, but only the idols that are in them. As for the buildings, just sprinkle them with holy water and place your altars and relics on them" (Choay 2017, p. 36).

About this reuse of the mosques, Choay (2017) reinforced that the monuments were reused because of economic reasons, which fit in the case of Seville. In this period, in times of crisis, the recommendation of the clergy was to maintain the buildings.

As for the cathedral church of Santa María of Seville, of the Gothic style, it was developed on the solarium of the old mosque of Aljama (Almohade period) in 1403. From the primitive temple, only the courtyard of "de los Naranjos" and its minaret are preserved. The Cathedral is an example of palimpsest, as in archaeological studies Roman and Moorish remains were found. Its bell tower, the Giralda, was the former minaret of the mosque, a masterpiece of the Almohade architecture. In the Renaissance period, Hernán Ruiz finished the Christian tower with bells (Tabales; Alba 2010). They went through other architectural styles such as: Baroque (1618-1758) and Neo-Gothic.

Another reuse was in relation to the Real Álcazar consisting of several palaces and gardens, which, after the Christian Reconquest, was used as accommodation for several monarchs and is used until now with this original function. On the Moorish monument, other buildings were constructed associating other architectures (Renaissance to Neoclassical) with the Moorish architecture.

From the fifteenth century, with the European commercial exploitation and consequently the knowledge of the new continent by Christopher Columbus (Cristóbal Colón), the city of Seville became a universal port, thanks to the Guadalquivir River. The city became very important (Fig. 2). The Catholic kings adopted the port of the city as the only route to America, since it was more protected than other port cities like Huelva and Cádiz. The kings were entitled to a commercial monopoly and established in 1584, the Casa Lonja or Casa de Contratación (Currently, General Archives of the Indies) to manage and control the gold and silver trade. It was built because there was a need for a suitable place for the merchants, since marketing near the Cathedral was generating conflict with the Church.

Its port was located next to the Golden Tower (Seville). At that time, the margins of the Guadalquivir River were sandy with great activity of people, travelers, merchants, and enslaved people. Between the years 1540 and 1550, Seville was the financial nucleus of Europe. This economic prosperity fostered arts, mainly the Sevillian Baroque (Murillo and Valdés Leal, among others), as well as financing the religious works, thus the city started to have numerous palaces (de las Dueñas, Casa Pilatos, etc.).

Within this context, Triana was responsible for the formation of sailors, which trained many of the crews for the Indies and America. In 1556, the University of Mareantes (now the monument of Casa de las Columnas) was founded. Still at that time, craft factories of shipyard, soaps (Almonas Reales), and expansion of pottery and ceramics were developed. Seville had a great urban expansion with about 100,000 inhabitants and was a world famous city.

In addition to training sailors, Triana was also responsible for repairing boats and supplying them with food production. The neighborhood was the largest of the extramural neighborhoods with 15,120 inhabitants (year of 1588), according to the ecclesiastical census cited by Ortega (2006). Their houses were modest and had many taverns. Besides the sailors, there were potters/ceramists, fishermen, and Portuguese residents.

The art and the technique of pottery/ceramics are traditional in Triana. The clays were extracted from the banks of the Guadalquivir River and transported by donkeys to the potters. This clay is very malleable and rich for art activities. They were also mixed with Aljarafe clays, resulting in very good material. The final results were very rich and unique painted ceramics. The historic relationship between ceramics and the Guadalquivir River landscape is like a fingerprint of Triana.

According to the Urbanism Management (Gerencia de Urbanismo, 1999), the sailors were close to the San Telmo bridge, on Betis/Pureza streets, near the port of Camaronero. Potters were in Barrionuevo (between Castilla and Alfarería streets); fishermen on the second stretch of the Rodrigo Triana and Convento Victoria streets, and the Portuguese merchants on the Cava street.

Focusing on the landscape representation of Triana (Fig. 2.1), we note densification of the houses and the beginning of an urban configuration, which in the future would be the San Jacinto, San Jorge, Castilla, Betis, and Pureza streets. Many trees can be seen near the fertile plain with native plants and vegetable gardens (current Pagés del Corro street or Cava street). People are observed using the space for leisure, conversation, and work. The Church of Santa Ana and the Castle of San Jorge are also present in this image. The Altozano of Triana was very important because travelers from Portugal and Alzarafe arrived there to do business with their agricultural products, there was the supply market, which later went to Seville though access to the Bridge of the Barges since the Almohade period.

Fig. 2 Seville in 1585. The urban growth of both Seville and Triana can be seen. *Source* View of Triana. Recorded by Ambrosio Brambilla, 1585. National Library of Madrid. It is present in the General Archive of the Indies, Seville. Author's photograph in 2017

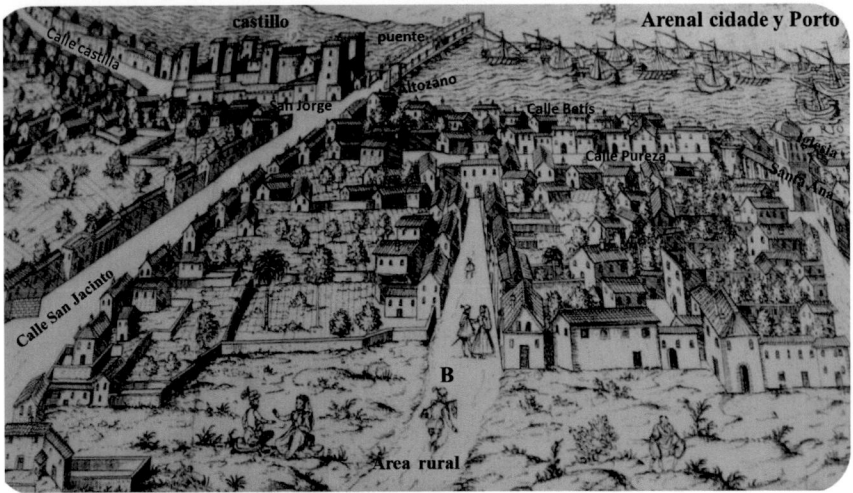

Fig. 2.1 View of Triana – 1585. *Source* Recorded by Ambrosio Brambilla, 1585. National Library of Madrid. It is present in the General Archive of the Indies, Seville, 2017. Author's photograph, including the names of the current streets in the historical neighborhood

With the colonial crisis, associated with the plague epidemic that devastated Seville and Triana in 1649 and the transference of the monopoly of the Indies to Cádiz in 1717, the city declined economically and socially. In the case of Triana, there were many epidemics, decimating the population. Ortega (2006) stated that, despite the demographic decline, Triana maintained its urban orientation.

Still, in the seventeenth century, Seville started to have several convents and monasteries of several orders (Franciscans, Jesuits, etc.).

In the eighteenth century, around 1705, the first descriptions of houses of collective neighbors. In Seville, they are popularly called "corrales de vecinos". They are properties of collective and modest housing around a central courtyard (Fernández-Salinas, 2003), where several families lived, appeared, mainly occupied by gypsies. According to Mantero (2015), at the time of King Carlos III, there were 1,000 gypsies in Seville (mainly in Triana on the Cava street) and many of them worked in the Royal Tobacco Factory (now the University of Seville). It is worth mentioning that Cava Street in Triana was a place where the flamenco was created, with songs linked to gypsy ethnicity.

Flamenco is one of the most artistic expressions of Andalusia. It is a fusion of vocal music, dance, and musical accompaniment and Triana was a singular center of creation. The root of this art is rooted in the singing of gypsies dedicated to ceramics and blacksmithing in Cava Street, in the middle of the eighteenth century. Flamenco was present in old taverns and parties at neighbors' houses.

In Seville, another culture that stands out is bullfighting. In 1707, there was a square bullring, which was renovated into a larger square—the "plaza de toros de la Maestranza".

In 1755, an earthquake caused several damages to the city, and in Triana, the Church of Santa Ana had to be rebuilt. Another problem faced by the neighborhood—the periodic flooding of the Guadalquivir River, had its solution at the end of the eighteenth century with the construction of a retaining wall on Betis Street.

In the nineteenth century in Triana, the church of the Chapel of Patrocínio and the chapel of the sailors (1805) were built. The immaterial values present in these churches and brotherhoods remain strong.

Between February 1810 and August 1812, French troops invaded Seville and Napoleón Bonaparte deposed King Fernando VII, and established his brother José Bonaparte as king of Spain (1808–1813). Triana stood out in the independence war, since the combat site was the bridge of Triana.

With the return of King Fernando VII, Seville was in a precarious state, as well as the rest of Spain, as a result of the independence process of the Spanish colonies. However, at the end of his reign, some urban changes occurred, continuing during the reign of Isabel II, who transformed Seville into a province in 1833; the arrival of the railway in 1859, in the city, among other transformations.

There were many changes in Triana: construction of a market to substitute the Castle of San Jorge (1825) and the construction of the new iron bridge Isabel II (1852), which will mean a new relationship between Triana and Seville, with adequate road connections.

Seville experienced the Spanish Revolution of 1868, where Queen Elizabeth II was dethroned and exiled, ushering in a period of instability. In Triana, the chapel of the sailors was expropriated by the revolutionary government in 1868. The images were transferred to a house and then to the church of San Jacinto (until 1962). On the same data, the wall (of the Moorish period) in Seville and the trace of the historic round were demolished.

At the end of the nineteenth century, an industrial period was inaugurated in Seville. Triana counts on the presence of artisanal activities and first factories, railroad, improvements in the port, increase in the neighborhood, and emergence of new neighborhoods with sanitation and urban planning projects. Many of these activities extrapolated the historic neighborhood of Triana. The neighborhood was very populous and with poor sanitation conditions, vulnerable to epidemics. Thus, the city hall started a sanitation project in 1859, on Cava Street, because there was a lot of garbage and standing water. This street became propitious to population growth.

In 1869, the City Council initiated the first alienations inspired by hygienist models, tearing down buildings and houses in the Altozano square and San Jorge and San Jacinto streets (Ortega 2006).

In the first half of the twentieth century, there were 53 industrial establishments in Triana. This first third of the twentieth century, provided work and demographic growth, interior renovation, new neighborhoods, streets, and a construction plan for the port against the 1900 floods, transforming the urban course of the Guadalquivir River into a hydrographic basin.

In 1914, the city of Seville inaugurated the María Luisa Park, and in 1928, the Spain Square is opened for the 1929 Ibero-American Exposition. In Triana, a hotel (the Triana Hotel) was inaugurated to be used by the guests of the exposition (IAPH 2017).

In Triana, there were a considerable number of immigrants, who occupied the houses and slums of the neighborhood, outside the historic Cava Street or Pagés del Corro Street. Therefore, the demographic increase caused changes in the landscape, originating neighborhoods outside the original nucleus, as already mentioned. Over time, the plains and vegetable gardens disappeared. According to Ortega (2006), this urban growth contributed to the ruin of part of the traditional houses.

In general, the Spanish economy developed in the late nineteenth century and early twentieth century, however, without many changes in the social conditions. In all the regions of Spain, anarchist and socialist workers rebelled against the "coup d'etat" of 1931. Triana is once again highlighted by the organization, but defeated by the harsh repression. At the end of this Spanish conflict from 1931 to 1939, the general Francisco Franco (1939–1975) established a fascist regime in Spain.

In Seville, the military initiated several public work projects. In Triana, the works for the installation of the tram ("tranvía") affected its interior, tearing down several buildings, such as the original Carmen chapel, the clock tower, etc. (Ortega 2006).

In 1956, the Interior Reform Plan of Triana led to the expulsion of residents, including the gypsies and destroyed several houses in the neighborhood. In addition, the own evolution of the buildings, in the name of the new industrial and commercial bourgeoisie of the neighborhood, was modifying and renovating the houses.

With the transition to democracy in Spain and the establishment of the Autonomous Community of Andalusia in 1981, with Seville being its capital, new paths for the country emerged with urban, social, and economic improvements.

In 1982, the revision of the Interior Reform Plan of Triana was carried out in order to establish protection for important buildings, approve management proposals on urban structure and management problems and improve the public facilities of the neighborhood. Consequently, many transformations related to infrastructure occurred in the city as a result of the Universal Exhibition of 1992, bringing new dynamics to Seville and Triana.

From 1990 onwards, the old and historic city of Triana was included in the Historical Set of Seville, with the aim to protect the historical monuments in order to prevent their disappearance using a Special Protection Plan. The Plan, approved in 1999, contributed to slow down the degradation process of the neighborhood, but it is a space in which capitalist interests are confronted, at all times, with public interests. The asset protection also acquired an appreciation of the neighborhood by property recovery companies, which elevated the property prices.

As for tourism activity, Seville was boosted since the beginning of the twentieth century, but, in fact, consolidated itself economically in the 1990s, with the Universal Exhibition. This event helped to revalue the historic center and further stimulated the cultural tourism. Since then, touristic offers and establishments have increased as a way out of the crisis of 2007. Currently, the city is breaking visitation records (2.6 million visitors in 2017), however, much is debated about the negative consequences of the activity.

In this context, the following items will present the historical and cultural heritage of Seville, as well as the threats arising from tourism.

3   Historical and Cultural Heritage of Seville

Before the fifteenth century, from what Choay called the "Italian Quattrocento", the interest in antiques, monuments was in the memorial sense, since the very meaning of the word monument comes […] "from the Latin *monumentum*, which in turn derives from monere (warn, remember), that which brings to mind something" (Choay 2017, p. 17).

This author explained that these buildings or antiques were preserved, not as an idea of heritage in itself, but for practical reasons of economy such as literary knowledge and sensitivity: "Intellectual charm, of course, but also sensitivity: the old works fascinate due to their dimensions, refinement and mastery of its execution" (Choay 2017, p. 37). The interest in monuments and antiquities arose around 1420, in Rome (Italy), at the moment when the object of Antiquity (mainly Greek) started to have value for its history, art, and aesthetics.

According to Choay (2017), the adjective as a heritage and historical monument occurred after the French revolution, when the distribution of the patrimony of the church, nobility, and the protection of another patrimony within the scope of the State was organized. Sapiezinka (2008, p. 1) corroborates the "patrimony transmitted from the Church to the people as a form of inheritance brought together real estate, monuments and statues and represented the public and collective property, laying the foundations for the construction of the notion of the national heritage". For Choay (2017) in the Commission for Historical Monuments of France, which appeared in 1837, there were three categories of monuments, such as remnants of antiquity, religious buildings from the Middle Ages and some castles.

According to Hobsbawn (2017), the production of traditions in Europe began at the end of the nineteenth century due to profound social transformations of the period, such as the emergence of Nation-State and the consolidation of the bourgeoisie. It was necessary to invent traditions to legitimize the nationalism and the economic/political power of the new social class and to manipulate the workers.

Hobsbawn and Ranger (2017, p. 342/343) report three novelties of the invention of the traditions. The first was the creation of primary education, the second the invention of public ceremonies, and the third the mass production of public monuments.

The cult of the historical monument took place around the 1960s, more precisely with the Venice Charter in 1964. The universalization of values expanded with the 1972 Natural and Cultural Heritage Convention (Choay 2017).

Over time, heritage typologies have diversified and there is the inclusion of the word culture, also disseminated in the 1960s. Monuments and other typologies become cultural heritage and cultural products.

As a product, monuments are consumed by cultural tourism since 1970s. Canclini (1993, p. 34) called this cultural consumption, that is "the set of appropriation processes and uses of products in which the symbolic value prevails over values of use and exchange, or when at least these latter are configured as subordinate to the symbolic dimension".

Cultural tourism "is at the origin of the expansion, perhaps the most significant, of the public of historical monuments—groups of initiated, experts and scholars were succeeded by a group on a world scale, an audience counting by the millions" (Choay 2017, p. 210). Later on, we will observe the impacts of this audience.

Regarding the heritages of Seville, it can be noted from Fig. 3 that the heritage of the monuments of the city began in the 1910s. By the end of the 1950s, there were only 18 heritage sites. In the 1960s, it can be seen that the number of listed assets is considerable when compared to the data from the 1970s. This is due, as Choay (2017) stated, to the fact that the monument cult gains strength in this decade, mainly because of the Venice Charter in 1964. From the 1980s, there is a resumption of the quantitative of the 1960s with the appearance of the category of historical garden and Seville gained a world heritage. From the 1990s onwards, new typologies appear, added to the quantitative increase in patrimonial assets until the year 2016 (Fig. 3). As a trend of the previously mentioned references, from the 1910s to 2016, the heritage assets of Seville were exclusively monuments with almost 80% (Fig. 3.1). Monuments, according to the Andalusian law, mean:

[…] buildings and structures of relevant historical, archaeological, paleontological, artistic, ethnological, industrial, scientific, social or technical interest with inclusion of furniture, installations and accessories that are expressly indicated (Art. 26 of law 14/2007).

From these monuments, the legitimation of religious, nobility, and bourgeoisie power can be observed, i.e., the dominant culture of different times.

The first monuments that became heritage date from 1912, and were exclusively religious buildings: the Chapel of San José, the Church of Santa Catalina, and in 1928, the Cathedral of Santa María (the Seville Cathedral).

After the Ibero-American Exhibition of 1929 in Seville, in the 1930s to 1939, twelve monuments were registered, mostly old convents and monasteries, old hospitals, churches (in Triana, the Church of Santa Ana, 1931), and palaces (Real Alcázar, Palace of las Dueñas (both from 1931).

Fig. 3 Number of heritage assets of Seville from 1910 to 2016. *Source* General catalog of the Andalusia historical heritage of the Culture Council of the Junta de Andalusia (Consejo de Cultura de la Junta de Andalucía, 2016). Organized by the author

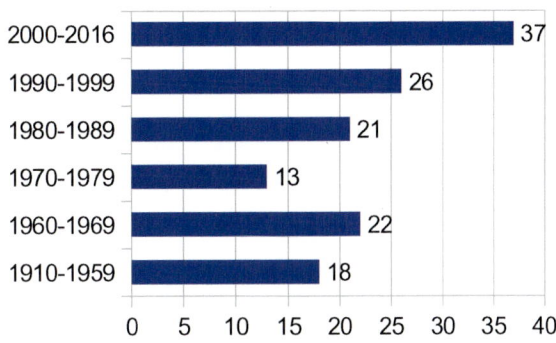

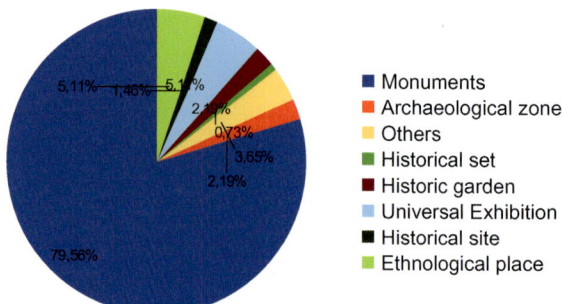

Fig. 3.1 Heritage typologies of Seville from 1910 to 2016. *Source* General catalog of the Andalusia historical heritage of the Culture Council of the Junta de Andalusia (Consejo de Cultura de la Junta de Andalucía, 2016). Organized by the author

After the Second World War, until the end of the 1950s, there were only three records of monuments (houses, churches) highlighting the Royal Tobacco Factory (now the University of Seville).

In the 1960s, there were 22 registers of monuments, most of them were old convents, old factories, churches, palaces, museums, and buildings related to military history. In the 1970s, the main monuments were old hospitals, old monasteries, military buildings, towers, etc. and the bridge of Triana (1976).

With the transition and return of democracy (Franco's dictatorship was from 1939 to 1975) from 1980 onwards, the registration of cultural assets were, in their majority, monuments such as old convents, palaces, Castle of San Jorge in Triana (1985), the "old lonja house" (today the Archives of the Indies), churches; however, there is the inclusion of the category of historical gardens (María Luisa Park). The Plaza de España, Real Maestranza bullring, museums (Murillo, art, and customs), and urban wall are noteworthy.

In 1987, the Seville/Giralda Cathedral, the Real Alcázar, and the Archives of the Indies became the cultural heritage of Mankind. A monumental complex.

For this, these patrimonies were recontextualized, and those that had individual values started to have universal values. For Veschambre (2008), this universal postulation, consensual, hides questions of social legitimation. Di Giovine (2018, p. 3) stated that even though "[…] each site was different, the UNESCO process would destroy their meanings and recontextualize them in an utopian metanarrative of unity in the diversity".

In 1990, the historical center of Seville was categorized as a historical set and there was the inclusion of the category place of ethnological interest with the registration of the Corral de la Encarnación in 1995, in Triana, patrimonies of the archaeological zone typology and churches and palaces like Pilates house appear. In Triana, there is the protection of the House of Columns (1990) and the church of San Jacinto (1990). The 1990s were also important, since Seville hosted, in 1992, the Universal Exhibition.

From the 2000s onwards, besides monuments such as old convents, palaces, Hotel Triana (2007) and two listed churches, a variety of legal typologies emerged, such as historical sites, historical gardens, and places of ethnological interest, like the Glass Factory of La Trinidad (2001), Corral de San José (2003), Escuela Sevillana de Baile

(2012), Escuela Bolera de Baile (2012), and La Carbonería, activity of Carbonería (2016).

It is observed that from the 2000s onwards, an evolution in the issue of immaterial culture began, related to the own evolution of the heritage debate, considering alternative cultures, as we saw with the typology of ethnological interest place.

Specifically about Triana, as the old part of the neighborhood integrated the historical set of Seville in 1990, it has become a protected landscape with heritages classified as cultural assets.

Most part of the Triana's heritage cited with full protection (Fig. 4) refers to the typology of the monuments, legitimized by the power of the Church and by the economic and political elites from different periods, such as the Church of Santa Ana, House of Columns, Castle of San Jorge, Church of San Jacinto, Bridge of Isabel II, and Hotel Triana.

The Church of Santa Ana has Gothic-Mudejar style and was built in 1266, after the Christian Reconquest. It was registered as a historical-artistic monument in 1931 and registered as a property of cultural interest in 1985 (IAPH 2017).

The House of Columns is a property registered in the category of the monument in 1990. It is a Patio House from the late eighteenth century (1780) in a Seville Baroque style. In the past, it was the Old University of Mareantes (IAPH 2017).

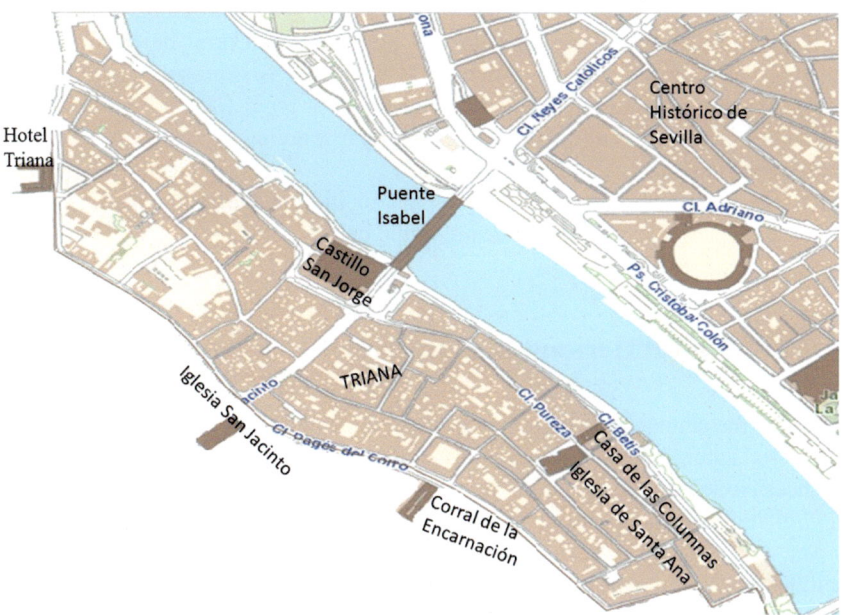

Fig. 4 Location of the assets of cultural interest in Triana—Seville (Spain), 2017. *Source* IAPH, 2017. Guía Digital del Patrimonio Cultural—http://www.iaph.es/localizador-cartografico-patrimonio-cultural-andalucia/lc_busqueda.php?municipio=6148

The law of 1949, which protects all Spanish castles, protected the Castle of San Jorge, in addition, since it is a property registered in the Spanish law of 1985. In the early nineteenth century, it was demolished to build a market, leaving only rubble and remains. Today, there is a museum of tolerance in its place (IAPH 2017).

The Church of San Jacinto is outside the historical set of Seville, but "there are parcels included around it that are within the scope of protection" (Urban Management 1999, p. 87). This church was registered as a monument category in 1990. The temple was opened in 1774, but it replaced an old one.

The Bridge of Isabel II (or Triana Bridge) is a property registered as a monument since 1976. The Triana Bridge came to replace the old Bridge of the Barges. It is an iron bridge of the nineteenth century, opened in 1852, built by the French engineers Bernadet and Steinacher (IAPH 2017).

Hotel Triana is a property registered as a monument in 1982. The building dates back to 1929, designed to serve as a hotel during the Ibero-American Exhibition of 1929, but it did not fulfill its function and ended up becoming houses. In 1981, the building was in ruins and the City Council bought it in 1983, saving the building. Currently, it comprises a group of houses (IAPH 2017). In the place, there are signs indicating that several artists lived there such as Gracia de Triana, Manolo Oliver (one of the great interpreters of Soleares de Triana), and Francisco Palacios ("El Pali"). The Flamenco Biennial takes place there and it is the headquarters of the North Triana Neighborhood Association (site 9).

The only typology of the ethnological place is the "Corral de Encarnación", showing the trend of change in the world debate. It is a property registered in 1995, for its ethnological value, representative of a typical house and traditional residents, configured as a form of collective life. It was constructed on the Chapel of La Encarnación.

The heritages of Triana are not limited to the cited assets. There is a rich ethnological heritage, which will be presented below.

4 Ethnological Heritage and Traditions of Triana

In Triana, the ethnological heritage expresses the culture and way of life, mainly of immaterial nature, related to symbolic and identity values, collective memories, and practices passed from generation to generation (Chart 1). Residents are resisting and legitimizing their traditions, giving meaning to the historical neighborhood.

The ethnological heritage corresponds to "the places, spaces, constructions or facilities linked to forms of life, culture, activities and modes of production typical of the Andalusia community" (Art. 61 of the Andalusian law 14/2007). The typology allows associating the activity of ethnological interest to movable assets and territorial areas by applying the protection regime corresponding to the activity.

The places that were valued in the neighborhood and chosen by members of the Northern Triana were the Guadalquivir River (with the Bridge of Isabel II), the

	Types	Place/ Examples.
Ethnological	Traditional knowledge, ways of doing work and crafts.	Ceramics. Ceramics Museum of Triana. Abandoned buildings of old factories. Pottery shops.
Ethnological	Traditional religious festivals	Holy Week, Corpus Christi, Romería del Rocío, Santiago and Santa Ana festivities, etc.
Ethnological	Oral Tradition	Legends, tales, poetry, music.
Ethnological	Typical houses	Patio Houses, collective houses (Corrales de Vecinos).
Ethnological	Movable ethnological assets	Religious images, mantles, embroidery, ornaments, etc.
Ethnological	Religious practices - Brotherhoods	Brotherhood of La Estrella (year 1566) from the La Estrella Chapel. Brotherhood of Hope of Triana (year 1418). Chapel of the Sailors Brotherhood of El Cachorro (year 1689). Church of Patrocinio, Basilica Christ of the Expiration - Brotherhood of the Parish of La O (year 1566). The brotherhood is older than the church. The church was built in 1702 on top of an old hospital. Brotherhoods of the Church of Santa Ana.
Immaterial	Flamenco Declared and intangible cultural heritage by UNESCO in 2010.	Origins: Cava Street (Pagés del Corro) by gypsies. Today, it is preserved in the Clubs, Dance Academies, Flamenco Biennial, flamenco dancers and singers, cultural events.

Chart 1 Ethnological and immaterial heritage of Triana. *Source* Participative observation of the author

Church of Santa Ana (including the square), Parish of La O, and the collective houses (Corrales de Vecinos).

The presence of the historical river is an identification and memorial of the maritime history and of the ceramic activity. Today, the river serves as a sport, leisure, and tourism.

Asked what it means to be from Triana, the interviewee said something about belonging. One interviewee said that being from Triana is not only just belonging to a neighborhood special because of its monuments, but because of the existing connections and conviviality with people:

> It is to feel that you belong to a special neighborhood; I do not mean only its cultural, architectural, artistic heritage, etc.… For me, where its true magic lies is because you can connect, that is, communicate in a more authentic and plain way with many of the people of your closest environment (associate of the Northern Triana Neighborhood Association, 41 years old).

The strong traditions are the religious practices and festivals. It is agreed with Veschambre (2008), that these performing celebrations are symbolic strategies of visibility and legitimation to mark the presence in a place. They constitute memorial landscapes through the performances of the bodies, and can be places of celebration of political expression, as stated by Alderman et al. (2020, p. 44):

> The metaphor of performance directs our attention to the ways in which memorial land-
> scapes serve as a stage, literally and figuratively, for a wide range of performances such as
> public dramas, rituals, historical reenactments, marches and protests, pageants, civic cere-
> monies, and festivals. It is not just that these performances happen in or at places of memory.
> Rather, the memorial landscape is constituted, shaped, and made important through the
> bodily performance and display of collective memories.

These religious performances in Triana are practices that unite the residents of
the neighborhood for their feelings, memories, and experiences, generating a feeling
of belonging to the place. Alderman et al. (2020, p. 45) present […] "that part of
what contributes to the increasing presence and significance of memorial making
and commemoration is the elevated place of emotion in contemporary society".

Among the religious practices, Holy Week expresses this emotion and commotion
on the streets and squares of the city. It is a party of international tourist interest (since
1980). The brotherhoods of all the city of Seville make their way to the Cathedral of
Seville with penance stations. The emotions of these processions are related to old
histories and traditions, which must be known to understand the symbolism.

Currently, flamenco has been reframed. That flamenco of the past was left behind,
since it was associated with the gypsies, who were expelled. Today, flamenco is
maintained in other ways (see Chart 1).

Thus, the ethnological heritages, mainly the religious practices and festivals and
everything that is related to them, reveal a strong feeling of belonging of the residents
in the neighborhood. As long as there are Triana people, embodying these experiences, creating and giving meaning to the rituals, their heritages, and traditions will
remain alive. However, we are concerned about the pressures of tourism in view of
the appreciation of the neighborhood.

5 The Most Visited Cultural Heritages and Tourism
in the Historic Center of Seville

Nowadays, tourism associated with the visitation of cultural heritage is a growing
economic activity in Europe, in the main cities such as Paris, Barcelona, and Madrid,
including Málaga, Granada, and Seville, which has had a negative impact.

In the Andalusia region (Spain), Málaga is the first place for tourism due to the
beaches and the weather, followed by other cities like Granada, Seville, Córdoba,
Úbeda, and Baeza, precisely because of their historical monuments.

In Seville, the most visited monuments were the Cathedral of Seville/Giralda,
followed by the Real Alcázar (Table 1). As it has already been mentioned, it was
declared a World Heritage Site by UNESCO in 1987. It can be seen that, geographically, the world heritage sites are concentrated, side-by-side, generating a crowd of
tourists at these points. Since UNESCO gave this title of World Heritage Site, there
has been an exponential increase in cultural tourism in Seville. As Choay (2017,
p. 211) said, the monuments "acquire a dual function – works that provide knowledge

Table 1 Most visited museums and monuments in Seville (Spain), in 2017 and 2018

Years	Cathedral of Seville	Real Alcázar	Palace Houses	Real Maestranza	Collegiate Church of the Divine Savior	Golden Tower Naval Museum	Mudéjar Center	Castle of San Jorge	Triana Ceramics Center
2018	2,123,836	1,875,771	383,412	362,019	341,652	185,692	41,926	32,673	27,304
2017	1,934,373	1,799,465	–	325,173	273,745	139,890	33,652	29,160	21,598

Source Consejo de Turismo de Sevilla. Centro de datos turísticos. Datos de Actividad Turística en la ciudad de Sevilha. Year 2018, p. 45. https://www.visitaSevilha.es/profesionales/actividad-turistica-de-Sevilha

and pleasure, made available to everyone; but also cultural products, manufactured, packaged and distributed to be consumed".

Di Giovine (2018, p. 7) mentioned that in the last decades UNESCO has sought greater involvement with the tourism sector. "This has led to a growing interest in better understand the effects of tourism, specifically in World Heritage sites, particularly in the ways that the local communities are impacted and involved.

Still on Table 1, in third place, in 2018, there was the route of the Palace Houses instead of the Real Maestranza, which in 2017, occupied this position. In eighth and ninth place are the museums in Triana—the museum of the Castle of San Jorge and the Triana Ceramics Center, which increased their visits by 12% and 26%, respectively, compared to 2017.

At the museum of the Castle of San Jorge, the tourists visit the Museum of Tolerance that shows the importance of respect and freedom of the people, besides showing archaeological excavations and historical collection. In Triana Ceramics Center, the museum tells the history of Triana, in relation to the traditional activity of pottery, heir to the Muslim techniques and Italian influences, as well as the history of flamenco. The museum is located in the old ceramic factory of Santa Ana (formerly Viuda de Gómez).

Tourists look for the image of Triana that has been sold to them: a neighborhood authentic for its ceramic history, modern (sailor), and cultural (flamenco). In the brochures and advertisements of tourism companies, the phrases "Triana is a neighborhood with essence" or "Triana is a picturesque neighborhood", are common.

The neighborhood was transformed. The gypsies expelled from the neighborhood, mainly by public politics of the twentieth century, left and with them the part of the essence of the neighborhood. There is no more pottery factory. The last one was closed in 2012. The factories were in decline due to the economic consequences of the Civil War and the entry of new architectural styles (Information obtained from the Triana Ceramics Museum). Nevertheless, it is the image of the past that permeates the imagination of the tourists. Choay (2017, p. 228) affirmed that the public visits these places looking for history and art, in fact, are deceived by the tourist industry that "tends to sell you illusions in the guise of the promised values".

Thus, tourists cross the Bridge of Isabel II to visit the historical neighborhood, which has a polysemic landscape for residents and tourists. From the observation decks (Colón or Golden Tower tour), tourists can admire the houses of Betis Street, the Guadalquivir River, and the Bridge of Isabel II that gives access to Triana (Fig. 5).

Its historical streets, its associated heritage, the market of Triana, the Altozano Square, and the river are important public identifiable spaces, valued by the residents, which often conflict with an excess of tourist audience who go to the neighborhood to know its monuments, its history, and festivities (mainly the Holy Week).

Tourists also visit other emblematic places of the neighborhood, besides the shops that sell objects and ceramic tiles. The flow is increasing every year.

In Seville, year-by-year, there is a series of concerns regarding the increase in the quantity of tourism. A study of Alonso and Tabales (2018, p. 372) affirmed that "the resident has full awareness of the strategic nature of tourist activity for the local economy and is willing to live with certain collateral effects, always within

Fig. 5 Houses of Betis Street and Bridge of Isabel II in Triana, from the view of the Golden Tower (Seville). Author's photograph, 2017

some boundaries". The residents pointed out the negative aspects of tourism, such as the complication of the access to the historic center, proliferation of commercial franchises, and touristification of the Santa Cruz neighborhood (Alonso; Tabales 2018, p. 373).

By touristification, the authors, Jover et al. (2017, p. 404), said that it is a process by which "the tourist activity symbolically and physically appropriates a territory. The mercantilization of the latter produces consequences for the people who inhabit it, changing its morphology, uses, identities, etc."

The newspaper "Diario de Sevilla"(2019) pointed out that the city is excessively economically dependent on the tourism section, even if there are good results, unemployment remains high in Seville.

Other serious problems arising from the tourism industry refer to the real estate market. Tourism is increasing rents and the price of real estate in tourist neighborhoods, causing expropriation of the residents. About the social exclusion, Choay (2017, p. 226) unveiled that:

> The packaging that is given to the urban historical heritage, in view of its cultural consumption, as well the fact that it is the target of investments of the prestigious real estate market, tends to exclude local or non-privileged populations from it and, with them, their traditional and modestly everyday activities.

Ortega (2006), who researched Triana, affirmed that the promotion of private housing over the public has allowed a free regulatory market with high prices. The rehabilitation of the houses developed a real estate promotion, since they buy and renovate older houses into modern apartments and sell them, immediately, at high prices. Regarding collective houses, the author draws attention that, despite the official help in the rehabilitation it is not effective, since it depends on the collaboration of the owners. There is pressure to tear down these traditional houses, declaring them in a state of ruins, increasing the eviction of the old residents of the neighborhood.

Another process generated from this promotion of the houses is gentrification, as studied by Días-Parra (2009), that is, processes of speculation of the soil that cause dynamics of ruins, evictions and new constructions, promoting and accelerating the process of expulsion of the lower classes of the area and the entrance of a population with greater purchasing power.

For one of the associates interviewed in the Northern Triana neighborhood, the folkloric model of the neighborhood is taking advantage of the social and environmental and is concerned with the real estate speculation and the collective sense, in the face of changes.

For an interviewee, it is important to "Intensify urgent messages about the heritage acquisition of the suburb", in the direction of (…) "a new model of neighborhood urge!" (A University professor, 56 years old). Another proposal for an associate is:

> Activities on weekends, of course, where older people can voluntarily explain the history of each of its corners with anecdotes about them, and vice versa also encourage young people and middle-aged people to participate in voluntary activities of empathy and collaboration with the elderly of the neighborhood (Associate, Aeronautical technician, 41 years old).

As general proposals for minimizing the negative impacts, generated by the tourist activity, the World Tourism Organization (OMT 2005) recommended the need for sustainable policies and management, incorporating economic strategies to the populations involved.

In addition, it is necessary to create debates involving universities, tourism companies, local companies, government, and civil society to carry out an inclusive heritage and tourism management, generating jobs and, above all, public policies to contain expropriations and social exclusion. It is necessary, urgent action to limit the number of visitors, because otherwise, as already indicated by Choay (2017, p. 228) "the exploration of the architectural historical heritage is, therefore, doomed to exhaustion, unless the maintenance costs are reduced and the flow of its consumers are regulated".

In 2018, the city of Seville discussed its II tourism plan with the participation of residents' associations, companies, and unions aiming at the sustainability (economic, social, and environmental) of tourism, ways to avoid the excessive concentration in the historical center and promotion of dialog with citizens regarding the tourist housing and necessary coexistence, avoiding tourism phobia.

In addition to these events, there are many others related to universities and organizations with experts in the area, contributing to the theme, as it was the case of the ESTAR event (Encuentro Social contra la Turistización: Alternativas y Resistencias, 2018), organized by the Cactus collective (Colectivo Asamblea contra la Turistización de Sevilla; see Barrero-Rescalvo 2019), formed by people and groups of the neighborhood associations, ecological movements, activists and academics. This seems to be the way, which requires a lot of collective effort.

It is essential to open a dialog with the residents about their views on the impacts of tourism, so that, together with the local government and other sectors, the activity be planned and territorially ordered, since when there are problems and the people are not heard, conflicts tend to increase in the face of touristification and the situation may be unsustainable.

6   Final Considerations

The studied landscape has several cultural heritage assets, chosen in the modernity and contemporaneity to be preserved. These monuments were legitimized by the dominant culture of each era, such as the Church, Christian kings, caliphate, and modern nobility, who managed to reproduce their values in the landscape. These monuments, such as the Cathedral, Real Alcázar, world heritage sites, are the most visited in Seville, reaching two million visitors.

At the end of the twentieth century, in the face of other paradigms, the ethnological heritage started to be recognized. When studying Triana, with regard to part of the historic center, it was noticed that even though the neighborhood has undergone transformations, the residents resist, that is, they claim and legitimize the place through their rites, their religious festivities, collectivity, dances, music, etc., giving value and emotion to the place. In fact, many of the monuments are connected to social values.

The appreciation of this historical heritage brought with it cultural tourism, as an economic alternative; however, it created arising mechanisms that compromise residents, who fight, among other questions, for the permanence in extremely disputed historical neighborhoods. Cultural tourism seeks experiences in traditional community events and festivities, as they have also become consumer goods.

The main problems arising from the tourist industry to the historic center refer to the excessive number of visitors, pressures of the real estate market with an appreciation of the properties, touristification, and gentrification process.

Thus, measures are needed to minimize the negative impacts of the tourist activity and it is essential to open a dialog with the residents about their views on the impacts of tourism, so that, together with the local government and other sectors, the activity can be planned and ordered territorially.

There are several aspects in this study that must be expanded and deepened, such as the possibilities of studies on the tourist imagination and a larger sample with residents to identify their perception of the impacts of tourism in their neighborhoods and in the city.

References

Alderman DH, Brasher JP, Dwyer III OJ (2020) Memorials and monuments. In: Kobayashi A (ed) International encyclopedia of human geography, 2nd edn, vol 9, Elsevier, pp 39–47. https://dx.doi.org/10.1016/B978-0-08-102295-5.10201-X

Alonso IM, Tabales AF (2018) Percepciones y valoraciones sociales del paisaje en destinos turísticos: análisis de la ciudad de Sevilla a través de técnicas de investigación cualitativas. Cuadernos de Turismo 42:355–383

Barrero-Rescalvo M (2019) Algo se muere de las Setas a la Alameda. Efectos del turismo sobre la población y el patrimonio en el casco norte de Sevilla. Actualidad, revista PH Instituto Andaluz del Patrimonio Histórico n.º 98 octubre, pp 46–49

Canclini NG (1993) El consumo cultural en México. Consejo Nacional para la Cultura y las Artes, México, p 1993

Centro de Estudios Paisaje y Territorio (CEPT) (2015) Catálogo de Paisajes de la provincia de Sevilha (Archivo digital)/directores: Florencio Zoido Naranjo and Jesús Rodríguez Rodríguez; autores: VV.AA.- Sevilla: Centro de Estudios Paisaje y Territorio, Consejería de Medio Ambiente y Ordenación del Territorio

Consejo de Turismo de Sevilla (2018) Centro de datos turísticos. Datos de Actividad Turística en la ciudad de Sevilla. Año 2018, p 45. Consorcio turismo de Sevilla. https://www.visitaSevilha. es/profesionales/actividad-turistica-de-Sevilha

Consejo de Cultura de la Junta de Andalucía (2016) Catálogo general del patrimonio histórico andaluz. Junta de Andalucía

Choay F (2017) A alegoria do patrimônio, 6th edn. UNESP, São Paulo

Cosgrove D (1998) A geografia está em toda parte: Cultura e simbolismo nas paisagens humanas. In: Corrêa RL, Rozendahl Z (orgs.). Paisagem, Tempo e Cultura. Rio de Janeiro: Eduerj, 1998, pp 92–122

Di Giovine MA (2018) O Patrimônio—Paisagem: Origens, Intervenções Teóricas e Recepção Crítica para a Compreensão do Programa do Patrimônio Mundial da UNESCO. Via Tourism Review, 13| 2018, posto online no dia 01 setembro 2018, consultado o 03 maio 2019. http://jou rnals.openedition.org/viatourism/2073, pp 1–12

Diario de Sevilla (2019) Los economistas advierten de la excesiva dependencia económica del turismo en Sevilla. Diario de Sevilla. https://www.diariodesevilla.es/sevilla/turismo-Sevilla-exc esiva-dependencia-economica-economistas-advierten_0_1361264148.html

Días-Garrido M (2004) Triana y la orilla derecha del Guadalquivir: evolución de una forma urbana desde sus orígenes hasta mediados del siglo XX. Unpublished doctoral thesis, Universidad de Sevilla

Dìas Parra I (2009) Procesos de gentrificación en Sevilla en la coyuntura reciente. Análisis comparado de tres sectores históricos: San Luis-Alameda, Triana y San Bernardo (2000–2006). Scripta Nova. Revista Electrónica de Geografía y Ciencias Sociales. Universidad de Barcelona, vol XIII, nº 304, Barcelona. http://www.ub.es/geocrit/sn/sn-304.htm

Fernández Salinas V (2003) La vivienda modesta y patrimonio cultural: Los corrales y patios de vecindad en el conjunto histórico de Sevilla. Scripta Nova. Revista electrónica de geografía y ciencias sociales, vol VII, núm. 146(070). Universidad de Barcelona, Barcelona. http://www.ub. es/geocrit/sn/sn-146(070).htm

Fernández SV (2008) El conjunto histórico de Sevilla: consideración, protección, amenazas y otros avatares. In: Grupo de Geografía Urbana (AGE): Guía para las excursiones del IX Coloquio y Jornadas de Campo de Geografía Urbana, Ceuta

Gerencia de Urbanismo (1999) Plan Especial de Protección del Sector 14. Triana, Gerencia de Urbanismo, Sevilha

Hobsbawn E (2017) A produção em massa de tradições: Europa, 1870 a 1914. In: Hobsbawn, E.; Ranger, T. (orgs). A invenção das tradições, 11th edn. Paz e Terra, Rio de Janeiro/São Paulo, pp 333–390

IAPH—Instituto Andaluz del patrimonio histórico (2017). Guía Digital del Patrimonio Cultural. http://www.iaph.es/localizador-cartografico-patrimonio-cultural-andalucia/lc_busqueda.php? municipio=6148

Jover J, Berraquero-Díaz L, Barrero-Rescalvo M, Jiménez-Talavera A (2018) Turistización y movimientos urbanos de resistencia: experiencias desde Sevilla. In: Mantero RS (2015) Historia breve de Sevilla. 2nd edn. Silex Ediciones, Sevilla

Nogué J (ed) (2008) El paisaje en la cultura contemporánea. Biblioteca Nuevap, Madrid

OMT (2005) Indicadores de desarrollo sostenible para los destinos turísticos: guía práctica. OMT, Madrid

Ortega JLR (2006) Triana y los Remedios durante el siglo XX: la conformación urbana del sector occidental de Sevilla. Diputación de Sevilla, Sevilla

Sapiezinkas A (2008) Do patrimônio histórico ao patrimônio cultural: diálogos e interações na aplicação das políticas públicas de preservação. Goiania, Habitus, vol 6, no 1/2, pp 67–101, jan./dez

Tabales MA, Alba M (2010) La ciudad sumergida: arqueología y paisaje histórico urbano de la ciudad de Sevilla (Proyecto Guía del paisaje histórico urbano de Sevilla). IAPH. https://reposi torio.iaph.es/handle/11532/326237

UNESCO (1992) Expert group on cultural landscapes. Guidelines on the inscription of specific types of properties on the World heritage list. La Petite Pierre, France

Veschambre V (2008) Traces et mémoires urbaines, enjeux sociaux de la patrimonialisation et de la destruction. Rennes, PUR

Rail Heritage and Museums

Museums, Railway Memories, and Cultural Landscapes

Davidson Panis Kaseker

Abstract In the state of São Paulo, despite its more than five thousand kilometers of railroads and hundreds of stations, warehouses, and residential complexes destined for railway workers, only 12 items are listed by the Instituto de Preservação do Patrimônio Histórico e Artístico Nacional (IPHAN) (Institute of National Historical and Artistic Heritage). In the list of the Council for the Defense of the Historical, Archaeological, Artistic and Tourist Heritage of the State of São Paulo (CONDEPHAAT), there are 32 railway buildings listed. The scarce number of railroad properties under the protection of the State's preservation agencies shows the selective criteria adopted by the official policies of historical heritage preservation. There are no official statistics, but it is known that many cities claim the historical value of their respective railroad heritage, taking into account the cultural relevance, usually associated with the origin of their own urban formations. Since the turn of the twenty-first century, the Brazilian and São Paulo railroad system has been in the headlines of newspapers and TV stations spreading images of abandoned and depleted stations, rolling stock depreciated by the weather and disuse, passenger cars, locomotives, and wagons rotting in urban yards. Theft of rails and poles became a routine agenda until the plundering was completed. The demand for the preservation and musealization of this type of heritage, stimulated by the bonds of affection in the imagination of a large part of the population that cultivates the railway memory, intensified in another way. The objective of this text is to point out guidelines for the musealization of open-air railway heritage, taking into account the methodological principles of museology and the distinct potentialities of cultural appropriation of these tangible and intangible assets by the populations of its surroundings with a view to the recognition of their cultural value and the regeneration of urban texture and social tessitura.

Keywords Memory · Railroad · Heritage · Musealization · Preservation · Brazil

D. P. Kaseker (✉)
Technical Coordination Group of the São Paulo State Museum System (SISEM-SP) Member of the Historical Archaeological, Artistic and Tourist Heritage Defense Council of São Paulo (CONDEPHAAT), São Paulo, Brazil
e-mail: dkaseker@sp.gov.br

Coming largely from the now extinct Rede Ferroviária Federal S/A—RFFSA (Federal Railway Network Anonymous Society)—Brazil's railway heritage is gigantic. The Brazilian state-owned railway company, created in 1957 and dissolved in 1999, was officially extinguished by Federal Law No. 11,483 of May 31, 2007, when it brought together 18 regional railroads operating in four of Brazil's five regions, with a presence in 19 federation units. In the State of São Paulo, the focus of this study, the RFFSA railroad network, inherited from Ferrovia Paulista S/A—Fepasa (Paulista Railway)—was 5,549 km long, formed by the extinct Sorocabana (2,016 km), Mogiana (1,744 km), São Paulo-Minas (133 km), Paulista (1,225 km), and Araraquara (431 km).[1]

Over the 50 years and 76 days of its existence, RFFSA has accumulated tangible assets, which can be subdivided into real estate and furniture, as well as intangible assets, which we will discuss later. Among the built patrimony are stations, warehouses, workshops, administrative buildings, villas, roundabouts, signaling booths, plots of land, yards, and even stretches of railway lines. In the category of movable assets, there are all kinds of rolling stock, such as locomotives, wagons, passenger cars, coastlines, machinery in general, in addition to three-dimensional collections, such as furniture, clocks, bells, and telegraphs. According to the inventory of the IPHAN, there are more than 52 thousand real estate and 15 thousand movable assets classified as of historical value by the Programa de Preservação do Patrimônio Ferroviário (PRESERFE) (Railway Heritage Preservation Program), developed by the Ministry of Transportation, the institution responsible for managing the mass in liquidation of RFFSA. In addition to the documentary collections, the properties belonging to the intangible heritage represented by cultural habits, technologies and knowledge produced by academic researchers, and memorials aficionados in the field of railway heritage are not found in this list.

The number of protected assets under patrimonial tutelage, therefore, does not by far show the expression that the railroad heritage finds in the collective imagination, especially with regard to the urban population. As for the documental heritage generated by the railroad, the collections made up of administrative documents, architectural and railway engineering projects, photographs, and others, including voluminous bibliographical collections, even though they are under the guardianship of official institutions, are still largely marginalized by public policies, reiterating a trend of official negligence present since the beginning of the first Republic. Indeed, in Brazil

> the word document, associated with the heritage and understood as an cultural heritage of historical value subject to protection, only appears for the first time in the 1946 Constitution. The 1967 Constitution, on the other hand, identifies documents as part of the set of cultural assets of the nation which, together with works and places of artistic or historical value, are under the protection of the State. And it was only in the 1988 Constitution that the national heritage received a more specific treatment (Oliveira 2019).

---

[1] The numbers are in the Wikipedia entry, available at https://pt.wikipedia.org/wiki/Ferrovia_Paulista_S/A. Accessed on 06/08/2019.

At the turn of the twenty-first century, due to the privatization process, the Brazilian railway system began to occupy the headlines of newspapers and television broadcasting images of abandoned and depleted stations, witnesses of the relapse with regard to rolling stock, passenger cars, locomotives, and wagons rotting in urban yards. In this step, without the due accountability of the perpetrators, crimes of prevarication and theft of rails, electrical wiring, and poles became routine until the plundering was completed.

As a response from civil society, the demand for the preservation and musealization of railway heritage has intensified in another way. The bonds of affectivity continue to mobilize the imagination of a large part of the population and that is why there are so many who cultivate the railway memory. In the university, interest has never waned, generating numerous investigations that address the history and development of the railroad as a driving force of São Paulo's economy in the pre-industrial period and, even today, relevant to the state's transport matrix. It is worth mentioning the Railroad Memory Project, coordinated by Ph.D. Eduardo Romero de Oliveira, from State University of São Paulo (UNESP), among so many other academic researches, with funding from the Foundation for Research Support of the State of São Paulo (FAPESP) and in partnership with the Public Archive of the State of São Paulo. The study promoted a quantitative inventory of the textual documentation of railroads in the state of São Paulo deposited in the collections of the Museum of the Paulista Company, in Jundiaí, in the State Archive, in São Paulo, and in the Memory Center of RFFSA, in Bauru. In general, these researches highlight the importance of the railroad for the creation and growth of cities and not infrequently point out the abandonment and forgetfulness of this heritage in the urban territory, which often makes it an open wound in the urban network, a vector that hinders the mobility of the population.

The abandonment and deterioration of railway heritage not only materialize as urban scars—in general the abandoned railway lines and stations have become targets of vandalism— but at the same time impact on the immaterial field the imagination of the population, disqualifying their self-esteem. Fact is that relations with the railroad inhabit to a lesser or greater degree the affective memory of each one, whether by personal experience, for generations who have worked in railroad companies or who have had the opportunity to travel in a passenger train, either directly or indirectly by connection with some family member, near or far, who is or has been a railroad. At the same time, from the point of view of the collective imagination, along with the nostalgic feeling, the railway memory is symbolically associated with the idea of progress and modernity to the point where the remaining steam locomotives are elevated to the category of icon of modern times. Feelings that are justified because they are historically rooted.

Ownership of railroads, ports, farms, warehouses and power plants - steam locomotives have witnessed an important part of Brazil's history in the long period from Mauá to the massive introduction of car transport. They were part of the economic cycles of the period, starting with coffee in São Paulo; they brought immigrants from the port of Santos to the Planalto Paulista; they moved sugarcane in the mills of the Northeast of Brazil and the North of Rio de Janeiro; they took and brought politicians and businessmen between Rio and São Paulo;

they made the iron deposits of Minas Gerais viable and introduced Brazil into the modern world (Perez 2019).

It is recurrent in the bibliography—notes Ana Lúcia Duarte Lanna—to highlight the relevance of the railroads in the expansion of the economy and in the occupation of regions of the interior of the country, through which the railroads transported people and goods, competing for the dissemination of cities and proliferation of regional centers. With the railroads, Brazil made the transition from an essentially rural country to a country with large urban populations. It is uncontroversial, in fact, to historically recognize the association of rail transport with the idea of progress and development, given

> the fundamental role that railroads have played in the growth and formation of cities, expanding the population and the demand for services and commerce, stimulating exchanges, shortening distances, altering the notion of temporality and introducing speed as a permanent feature of the ways of life established with the advent of capitalism (Lanna 2016).

It is also undeniable the influence that the architectural patterns of the cities received from the industrial heritage of the railways built in the late nineteenth and early twentieth centuries, with new techniques, new materials, and styles, often imported from other countries. Definitely, for good or evil, the railway heritage would alter the contours of urban spaces and change the *modus vivendi* of the populations and not only those around them.

With this, Lanna (2016) stresses, "the railroad materializes the triumph of the technique and incorporates it into everyday urban life, creates new characters and is essential in the organization of the world of work". The railroad will stimulate the development of production, which will later be called *agro-business* focused on the domestic market and exports. Beyond this association of the railroad with the economic and social development in Brazil, foreshadowing what would be understood as the "country of the future", Lanna highlights something that is fundamental to unveil the symbology created in the field of railroad memory. By being "indissolubly associated with speed, a great icon of the modernity of the nineteenth and twentieth centuries, railroads increase mobility, consolidate the practice of travel, are part of a new picture that is outlined in Brazilian society at the end of the nineteenth century, the *Belle Époque Tropical*" (Lanna 2016).

Not only glamor, however, is the history of the railroad written in Brazil, as it is all over the world. Quoting Wolfgang Schivelbusch, Lanna observes that the rail transportation system, unlike in Europe, where its implementation was marked by the destruction of traditional landscapes instituted by an established culture, in America "constitutes a way to conquer the inaccessible nature in favor of a civilization and economic production" (Lanna 2016).

From a critical perspective, therefore, it must be recognized that, to a large extent, at the extremes of São Paulo territory, driven by agricultural expansion based on the coffee culture, the railroads have paved the way for the devastation of forests, for the indigenous massacre, and the deterritorialization of dozens of remaining ethnic groups that still struggle to preserve their cultures and seek to give voice to the silent memories. The railroads, created without the public power having previously conceived a strategic planning that characterized an integrated transport system, developed taking into account the particular interests of rural producers and investors, in such a way that their routes did not always take into account "the geographical conditions of the sites, but mostly obeying the interests of politicians and companies that obtained the concession, which included land and operation" (Mazzoco 2005, *apud* Cruz 2007, p. 66).

By giving priority to the entrepreneurship of the "pioneers" of the occupation of São Paulo's territory and the fruits of economic development resulting from the implementation of the railroads, the narrative aligned with the viewpoint of the dominant power omits this other side of history and, as a rule, does not even recognize the erasing of the memory produced by it. Not for less, for the majority of the population, the affective memory of the railroad is mixed with nostalgic feelings of a mythical past, marked by hegemonic values of security, permanence, and stability that contradict reality. The railroad has thus become an icon of modernity, as has already been said, although the modernization of Brazilian society remains incomplete and inaccessible to the vast majority of Brazilians, still today denied the benefits associated with the celebrated progress, both in the area of transportation infrastructure, basic health and sanitation services, education and, especially, with regard to human rights and the environment.

In this sense, despite the relevance of the social memory erected from an imaginary that includes experiences from the railroad's past linked to dreams, illusions, affections, and feelings, as well as the railroad's contribution to the passage of rural Brazil to a predominantly urban and industrialized country, until the present, however, the preservation of this heritage and its resignification remain a challenge if we want to unveil their narrative gaps and lapses.

Part of this contemporary challenge is the exercise of some indispensable reflections: Who defines what deserves to be preserved and what should be memorable in this process of recognition of the railway heritage? Who is responsible for safeguarding the cultural heritage? How is it possible to safeguard the cultural heritage, be it tangible or intangible? And finally, how to mediate the cultural heritage of the railway?

The legal instrument that consolidates the creation of the preservationist policy in Brazil, through which the Serviço de Proteção do Patrimônio Histórico e Artístico Nacional (SPHAN) (National Historical and Artistic Heritage Service) was also created—later succeeded by the IPHAN—is Decree-Law No. 25, of November 30, 1937, in whose scope is located the legal instrument for heritage protection, as the administrative act in Brazil that aims to prevent the destruction, degradation, or mischaracterization of cultural goods recognized as such is called. Once registered in one of the four Books of the Tombo instituted by the Decree—Book of the Archaeological, Ethnographic and Landscape Tombo; Book of the Historical Tombo; Book of the Tombo of Fine Arts; and Book of the Tombo of Applied Arts, the property becomes protected by the State, even if its ownership continues to belong to its owners.

It should be noted that the Books of Tombo are an allusion to the Torre do Tombo (Tumble Tower) erected in Portugal during the reign of King Fernando in the fourteenth century, which was established as the public archive responsible for safeguarding royalty and public administration documents. In the eighteenth century, it physically collapsed with the earthquake that destroyed Lisbon, but its institutionality was preserved, and today the National Archives of Torre do Tombo are subordinated to the General Direction of Books, Archives, and Libraries. The word historically is associated with antagonistic meanings. Popularly, the tombstone is synonymous with overthrowing, demolishing, destroying, and, in legal terms, it has the opposite meaning in that it registers something that is of value to a community by protecting it through specific legislation.

The cultural heritage was then defined in Article 1 of Decree-Law No. 25/1937 as "a set of movable and immovable assets existing in the country and whose conservation is of public interest, either because of its attachment to memorable facts in the history of Brazil, or because of its exceptional archaeological or ethnographic, bibliographic or artistic value", or of natural landscapes of "remarkable feature" (Brazil 1988).

The first years of SPHAN's work were difficult. Besides the Herculean task of protecting an immense set of heritage assets dating back to the sixteenth century, there was still the challenge of building a work methodology, aligning concepts, and facing the dominant mentality that, in a period of accelerated urban expansion, associated progress with the destruction of the past, exposing in general the built heritage to the ravages of real estate speculation. At this stage, had it not been for the intellectual vigor and contribution of personalities such as Gustavo Capanema, Mário de Andrade, and Rodrigo de Melo Franco de Andrade, the battle would have been lost.

Anchored in the purpose of building a national identity in line with the ideological guidelines of the Vargas era (1930 to 1945), in this first phase SPHAN's actions favored the safeguarding of the heritage built in earthenware, stone, and lime linked to the colonial past and the Catholic Church, leaving as his greatest legacy the idea that it is possible to promote the transformation of the country without the destruction of the historical, artistic, architectural, and archaeological heritage.

In this line of action, however, assets that were not linked to memorable facts from the point of view of the official history of the country were excluded, although memorable for the understanding of neglected layers of this history. Also excluded by this monumentalist vision, which even today reverberated in certain segments of society, are those goods that do not present the desired exceptionality, despite their archaeological or ethnographic, bibliographic or artistic value for the local communities or even in the absence of the affective relationships that the population establishes with them. Thus, the understanding that the protection of cultural assets that did not contemplate the national interest would depend on complementary legislation in the state and municipal spheres, when and if they exist, was established.

To the valorization of national monuments, in this sense, it is opposed the pressure of intellectuals who defend the formulation of public policies aimed at the recognition of the cultural good as a living, changing and immaterial heritage, respecting the cultural production as peculiar processes inserted in their social contexts, with the aim of promoting the autonomous development of spontaneous cultural manifestations. The innovative ideas of Mário de Andrade and Paulo Duarte, which had been swallowed up by monumentalist appreciation, are taken up by Aloísio Magalhães in the 1970s with the creation of the Centro Nacional de Referência Cultural—CNRC (National Center of Cultural Reference). The understanding of the scope of the concept of cultural heritage is growing. In terms of preservation policy, the need to update the concept of preservation is gaining strength.

In the 1970s, under a military regime, as a result of the efforts of states and municipalities to engage in the preservation of historical monuments of recognized relevance and the preservation of the built heritage of the so-called historical cities, based on the understanding that cultural heritage is a factor of local and regional development through tourism, emerge from this historical context the ideas of decentralization and international influence of preservationist policies spread by the United Nations Educational, Scientific and Cultural Organization (UNESCO), especially the Heritage Charters. The Venice Charter (1964), the basic document of Icomos (International Council of Monuments and Sites), whose repercussion took a long time to be assimilated in Brazil, and the Cultural Tourism Charter (1976), among others, should be highlighted. The historian Ana Luiza Martins (2018, p. 60) also highlights that the Declaration of Amsterdam (1975) was decisive, which explicitly advocated the importance of preserving urban complexes of "ancient cities and villages with traditions in their natural or built environment" and not only the most important monuments.

From there, in the period that extends to the redemocratization of the country, an effort to professionalize the sector prevails, even if in a fragmentary way, with emphasis on the expansion of cultural access to heritage, the appreciation of cultural diversity, and the intensification of the debate around the selection criteria for the protection of cultural heritage.

As a reflection of the process of political openness, the new master charter of 1988 brings the concept of cultural law, strongly inspired by Article 37 of the Universal Declaration of Human Rights (1948), according to which "everyone has the right to take part freely in the cultural life of the community, to enjoy the arts and to participate in scientific progress and the benefits that result from it. In Article 216 of the Brazilian Constitution, the concept of cultural heritage effectively gains a wider scope:

Brazilian cultural heritage is made up of material and immaterial goods, taken individually or together, bearing reference to the identity, action, and memory of the different formative groups of Brazilian society, in which they are included:

I - the forms of expression;

II - the ways of creating, doing and living;

III - scientific, artistic and technological creations;

IV - the works, objects, documents, buildings and other spaces intended for artistic-cultural manifestations;

V - the urban ensembles and sites of historical, landscape, artistic, archaeological, paleontological, ecological and scientific value.

In this context, the so-called Citizen Constitution will broaden the horizons of cultural rights, as it is clear from its Article 215:

> The State shall ensure the full exercise of cultural rights and access to the sources of national culture for all, and shall support and encourage the enhancement and dissemination of cultural events.
>
> § Paragraph 1 - The State shall protect the manifestations of popular, indigenous and Afro-Brazilian cultures, and those of other groups participating in the national civilization process.

In these two articles, the 1988 Constitution "not only advocates rights, but also guarantees, and outlines instruments for their realization" (Barbosa 2015, pp. 73–104). In Article 215, the exercise of cultural rights and access to sources of culture are ensured; in the following, cultural production and its knowledge. Thus, it explicitly proclaims the action of public power to offer the conditions for the exercise of cultural rights, opening a participatory perspective as a method of implementing public policies, as is evident from Art. 216:

> §1° The public power, with the collaboration of the community, shall promote and protect the Brazilian cultural heritage, through inventories, registration, vigilance, protection and expropriation, and other forms of caution and preservation.

That is, for the first time the legislation recognizes the importance of sharing with society the responsibility for the preservation of cultural assets through popular participation in this process. In fact, besides the Union's own notorious difficulty in implementing preservationist public policies and the difficulties that states and municipalities have in recognizing and protecting their cultural assets through their own legislation, the effectiveness of the administrative act of protection should be questioned as to the ability to ensure the preservation of the cultural asset, if the official recognition, through the bodies responsible for protecting the heritage does not correspond to the engagement of society.

In this respect, the anthropologist and professor at Universidade de São Paulo (University of São Paulo), José Guilherme Magnani, warns:

> The so-called historical cities are not only scenarios of old events that still retain, in the layout and home, the marks of the time; it must be recognized that life, there, continues. The relations between the current actors with these scenarios, however, are not always taken into account by the preservation agencies. This omission can be perceived in some premises that guide preservationist practice. The first is the assumption that the criteria with which cultural assets are selected and classified are universal and that they are shared homogeneously by all users. The other is to consider the latter as mere obstacles to preservation, since most of the time the relationship between users and preservationist bodies is conflicting, both with regard to the criteria of choice and with regard to state intervention through the protection mechanism. (Magnani and Martins 2018, p. 92)

Nevertheless, all the difficulties faced in the field of cultural heritage preservation, another important milestone in the process of building the institutionality of public policies in the area of culture is Decree No. 3,551/2000, which institutes the Registration of Cultural Assets of Immaterial Nature, in the scope of which IPHAN created the National Program of Immaterial Heritage, with the objectives of emphasizing the duty and strategic centrality of the State in the documentation, registration, and inventory of cultural assets, giving visibility to the cultural diversity of the groups that form the Brazilian society, in addition to opportune the exercise of the right to memory and propitiate to the diverse social groups the claim of copyrights resulting from the use of traditional knowledge and doings.

The decree highlights the expansion and explanation of the field of culture in its multiplicity, dynamism, and plurality, reaffirming the references to the 1988 Constitution and bringing cultural policies closer to the anthropological–ethnographic concept present in the initial formulations proposed by Mário de Andrade. The cultural manifestations cease to be mere folkloric curiosity and become recognized as an affirmation of the intrinsic dignity of their practitioners as well as the social, artistic, and historical value they represent for the whole country.

Cultural rights, recognized by international treaties as part of second generation human rights, together with economic and social rights, are now also promoted here in Brazil by State action. Neither privilege nor favor: culture gains the scope of a right.

The internalization of this statute, however, is markedly precarious, even though it is appropriated by broad social segments. No wonder it is still common to hear that one should not spend on culture until health and education are fully attended to, in an obviously mistaken view not only of the indissociability of such issues but also of the very dichotomization between spending and investment.

In addition to this historical context, notwithstanding the protagonism exercised by the Union in the formulation of public policies in this field, the fact that cultural policies are competing competences, as defined in the 1988 Constitution, shared among federal, state, and municipal agencies, in addition to third sector institutions, makes it difficult and complex to cover the normative potential as well as the allocation of financial resources for the preservation of cultural heritage.

In this sense, since the 1988 Constitution, despite the fact that the programmatic efforts of the former MinC[2] were guided by the conceptual advance that recognizes culture as a fundamental right of the citizen, such initiatives still today find it difficult to sustain themselves as a state policy, "remaining subject to the back and forth of macroeconomic policies or to the idiosyncrasies of governments on duty, which hinders a radical and effective change towards a cultural policy understood as long-term development policy" (Bolaños et al. 2012).

Despite the relevance of the laws of fiscal incentive to culture, conceived as tools of legal and economic nature intended to ensure the preservation of the cultural assets of the country, in practice, in a democratic system in which the Provider State is emptied of its possibilities of action/intervention, several obstacles converge for the effective implementation of public policies, starting with the scarce budgetary resources made available for the structuring of public services in the preservationist bodies, the lack

---

[2]Created on March 15, 1985 by Decree No. 91,144 of President José Sarney, the Ministry of Culture (MinC) was transformed into the Secretariat of Culture on April 12, 1990, in the government of President Fernando Collor de Mello, recreated on November 19, 1992, by Law No. 8. On January 1, 2019, after the administrative reform of the newly appointed government of Jair Bolsonaro, the Ministry of Culture was officially extinguished by Provisional Measure No. 870, returning to the status of Special Secretariat.

of investment in scientific production and in the training of specialized professionals, and even the conflicts of the social forces and the persistent intrinsic incapacity of intra-institutional articulation of the public power itself.

As far as railway heritage is concerned, for example, several federal bodies compete for the administration of cultural goods that historically have had little dialogue with each other. The Secretariat of Union Patrimony (SPU) is in charge of the incorporation and regularization of the real estate property belonging to the Union, as well as its proper destination, in addition to the control and inspection of the referred assets. In turn, the same law that extinguished the RFFSA attributes to IPHAN the responsibility of receiving and managing its movable and immovable assets of cultural value, as well as taking care of their safekeeping and maintenance, being also responsible for evaluating, among all the assets originating from the extinct network, which are the assets with historical, artistic, and cultural value. In the division of duties among the Union bodies, non-operational assets are transferred to the Institute and other federal entities, while operational assets remain under the responsibility of the National Transportation Infrastructure Department (DNIT), which is also responsible for distinguishing, among unserviceable assets, those that eventually have historical value. In other words, a cipher of prerogatives and functions makes it difficult for the State to act in an articulated manner from the point of view of its operationalization.

Apart from the structural issue, with regard to the implementation of public policies for the preservation of the railway heritage, the size and diversity of this set of assets still weigh heavily. The Railroad Cultural Heritage List, created through IPHAN Ordinance No. 407/2010, includes thousands of movable and immovable assets of RFFSA registered until December 15, 2015. Official data indicates that in total there are 142,277 non-operational movable assets analyzed by IPHAN with an indication of transference to city halls or non-governmental organizations; 1,034 operational movable assets analyzed by IPHAN and transferred to DNIT; 27,697 non-operational movable assets analyzed and incorporated to IPHAN; and, in particular, 639 real estate assets valued and enrolled in the list of heritage assets to be preserved.

In the state of São Paulo, despite its more than five thousand kilometers of railroads and hundreds of stations, warehouses, and residential complexes for railway workers, only 12 properties are listed by IPHAN. It should be noted, however, that the immense majority of national heritage related to the Brazilian railroads made by the federal agency date from the last decade, reflecting a considerable effort to overcome these liabilities.

On the list of the Conselho de Defesa do Patrimônio Histórico, Arqueológico, Artístico e Turístico do Estado de São Paulo—CONDEPHAAT (Council for the Defense of the Historical, Archaeological, Artistic and Tourist Heritage of the State of São Paulo)— there are 32 properties listed that are directly or indirectly related to the railway memory. To systematize and also build the knowledge about the São Paulo railroad heritage, as provided in Article 64 of State Decree 50,941/2006, technicians[3] from the Historical Heritage Preservation Unit (UPPH), an instance of the Secretariat of Culture and Creative Economy of São Paulo, developed thematic studies based on a methodology that guides the selection of railway assets, which, taken individually or together, may be considered "bearers of references to the identity, action and memory of the different groups that form society", as Zagato quotes in a thematic study prepared by the UPPH, in accordance with the Federal Constitution of 1988 (Art. 216) and the State Constitution of 1989 (Art. 261).

The proposal aims "to allow the evaluation and comparative perspective assessment of each good based on objective but sufficiently broad criteria that ensure respect for its specificities", always bearing in mind state relevance. The UPPH study group— composed of historians and architects—focused on the data and covered preliminarily six of the nine original railway lines in São Paulo territory. The study group studied the São Paulo Railway (SPR)/Santos-Jundiaí Railroad (EFSJ), the Companhia Paulista de Estrada de Ferro—CPEF (Railway Company Paulista), the Estrada de Ferro Soro-cabana—EFS (Sorocabana Railway) including Ituana, the Companhia Mogiana de Estrada de Ferro—CMEF (Railway Company Mogiana), the São Paulo branch of the Estrada de Ferro Central do Brasil—EFCB (Central Brazil Railway), the Estrada de Ferro do Noroeste do Brasil—EFNOB (Northwest Brazil Railway), Estrada de Ferro de Araraquara—EFA (Araraquara Railway), and Estrada de Ferro Bragantina—EFB (Bragantina Railway), leaving the Estrada de Ferro de Campos do Jordão—EFCJ (Campos de Jordão Railway) (Martins 2018, p. 152).

Map of São Paulo Railway Lines

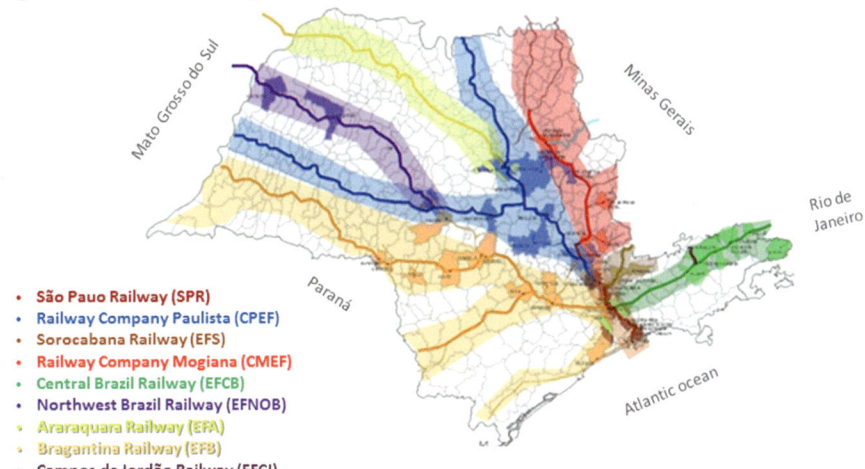

- São Pauo Railway (SPR)
- Railway Company Paulista (CPEF)
- Sorocabana Railway (EFS)
- Railway Company Mogiana (CMEF)
- Central Brazil Railway (EFCB)
- Northwest Brazil Railway (EFNOB)
- Araraquara Railway (EFA)
- Bragantina Railway (EFB)
- Campos de Jordão Railway (EFCJ)

Source UPPH

---

[3]The Railroad Project team is composed by architects José Antonio Chinelato Zagato (Coordinator; 2009–2019), Alberto Fernando A. Candido (2009–12), and Adda Ungaretti (2012) and historians Amanda W. Caporrino (2010–2011; 2018–2019) and Ana Luiza Martins (2008–2012). The group is subordinated to the director of the GEI Elisabete Mitiko and the coordinator of the UPPH Valéria Rossi Domingos.

They already have published Protection Resolutions, the Funicular System of Paranapiacaba (SPR), the railway complexes of Santos (SPR), Jundiaí (SPR), Botucatu (EFS), Louveira (CPEF), Sumaré (CPEF), Bauru (EFNOB, CPEF and EFS), and Sorocaba (EFS). In conclusion, there are other "expressive sets, such as the Mogiana Workshops (Campinas, CPEF), Guanabara (Campinas, EPEF), the Lapa workshops (São Paulo, SPR), Pátio do Pari (São Paulo, SPR) and Piracicaba (CPEF)" (Martins 2018, p. 153).

As criteria for the identification and selection of railway assets with protection potential, the UPPH contemplates (1) the main points of the railway lines [junctions, landmarks: terminal points (railheads), transpositions (mountains, rivers, etc.)]; (2) the railway industrial complex with favorable legibility, preferably intact or almost intact; (3) reference points for historical, social, and economic processes at regional, state, or national level; (4) Sets and/or works that present techniques, architecture, and/or engineering of special character; (5) the insertion in trunk lines and/or branches of regional/state economic, historical, or social expressiveness; (6) the operational class of the station; (7) the state of characterization and/or conservation favorable to preservation; and, as additional criteria, (8) the location in Municipalities of Tourist Interest (MIT).

According to data from the UPPH's Study Group for Inventory and Recognition of Cultural and Natural Heritage, by 2018 slightly less than 120 applications were received for the protection of railway national heritage, of which approximately 30% were listed as national heritage and homologated, another 10% are in the process of being homologated, the rest of the applications are still under study and slightly less than 40% were filed.

Zagato highlights the symbolic and pedagogical role of the state patrimony agency, which, in addition to the quantitative advances in the process for registration of the cultural heritage and, in the technical sphere, the qualification and rationalization of the processes of analysis submitted to the CONDEPHAAT, which has contributed to a progressive mobilization of society, the Public Ministry, and local governments, especially in the Interior of the State, "where the railway assets have even greater affective value, because they still mark presence in the urban and rural landscape" (Martins 2018, p. 153).

According to Zagato, many of these goods "have been preserved and converted into public equipment with various uses—from simple local memory centers to interesting museums, auditoriums, municipal administrative departments, schools, welfare centers, etc." (Martins 2018, p. 154).

There are no official statistics, but it is known that many cities claim the historical value of their respective railway assets, given the local relevance, generally associated with the origin of the urban formation itself or even because they are associated with the affective memory of their residents. No wonder there is such a social demand for the transformation of railway heritage into heritage, given its historical, artistic, and cultural legacy. The musealization of heritage assets, once recognized as indicators of railroad memory, has effectively been sought by resignifying these collections as potential guardians of social memory and, at one time, with the purpose of leveraging cultural tourism as an axis inducing local development.

In fact, mapped by the State System of Museums (SISEM-SP) in 2010, in São Paulo, there are six railway museums open to the public and 14 museums—most of them historical museums—based in old railway stations, in addition to hundreds of historical museums and other similar institutions that maintain and preserve railway assets—office furniture, clocks, telephone sets, telegraphs, bells, machines, and tools, including rolling stock, locomotives, and wagons. Other alternatives have also been undertaken for the resignification of the built railway heritage. In São Paulo, there are at least two dozen cultural spaces installed in former stations (Kaseker 2018, pp. 165–175).

As can be seen, there are different forms of cultural appropriation of railway memory. The creation of railway museums, in the classic format of institution as defined in Brazilian legislation by the Statute of Museums,[4] is only one of them and is not even the most usual, although it is the one with the highest social prestige. The guarding of collections of objects in places of memory or even the conservation of buildings representative of railway heritage are more commonly practiced. For these alternative practices of railroad heritage preservation, it is necessary to consider that in the last fifty years not only the concept of cultural heritage has expanded, but also the horizon of musealization has expanded.

In truth, throughout the twentieth century, not only did the number of museums worldwide multiply, following the growth of large cities, but also the profile of museums gained new contours. If before, still confined between four walls in European countries, the encyclopaedic museums, as initially conceived, are organized around collections, in the form of urban cultural equipment qualified to exercise the "representation of nationalist principles and as an expression of expansionist and colonizing policies, enshrined in the eighteenth and nineteenth centuries" (Bruno, M.C.O. 2000), the museums are gradually assuming themselves, worldwide, as active subjects in the path that leads to the preservation and administration of memory and cultural heritage as a whole.

The Santiago Round Table, held in Chile in 1972, is the frame of reference in the inflection curve of this trajectory of construction of a new scope of museums. Promoted by the International Council of Museums (ICOM), a non-governmental organization associated with UNESCO, then under the presidency of Hugues de Varine-Bohan, the Santiago Round Table would result in the conception of the integral museum, consecrating two orientations for the museums' action based on this concept: the social function of the museum as an instrument of social and cultural development at the service of a democratic society and the commitment to the integral heritage, bases for the social intervention in the territory's cultural and natural, tangible and intangible heritage.

Two contributions were acclaimed at the Santiago Table, which brought together the directors of the main Latin American museums. First, the thought of Paulo Freire,

---

[4]Cf. Federal Law 11.904/2019, Article 1: *For the purposes of this Law, museums are considered to be non-profit institutions that conserve, investigate, communicate, interpret, and exhibit, for preservation, study, research, education, contemplation, and tourism purposes, sets and collections of historical, artistic, scientific, technical, or of any other cultural nature, open to the public, at the service of society and its development.*

who even though prevented from attending the event by the Brazilian civil-military dictatorship, inspired the debates with his pedagogy of liberation, substantive in the valorization of transdisciplinarity that puts all knowledge on an equal footing, without hierarchizing them, as summarized in the phrase "there is no knowing more or knowing less. There are different knowledges" (Freire 1987, p. 68). Not least, the participation of the Argentine architect Jorge Hardoy deserves to be highlighted. He provoked the discussion of the importance of the museum, whatever its size or typology, to look at the reality of its surroundings and in some way contribute to its transformation, especially in the context of precarious living conditions in large urban agglomerations.

The inaugural landmark of a new museology, "the 1972 Santiago de Chile Round Table Declaration symbolically highlights the implosion of secular values, triggering a search for new paths for musealization processes", points out Cristina Bruno (2010). Although these innovative conceptions have not brought about major changes at once, not even in large Latin American museums, the principles of Santiago's museological pedagogy have taken root profoundly among museologists and museum professionals, as evaluated by Bruno:

> The concept of an integral museum questioned consecrated notions of the museological universe such as collectionism, the museum between four walls and the official heritage. It has awakened the attention of professionals to an entire heritage awaiting musealization, to the importance of community participation in all museological instances and has imposed new working methods (Bruno 2010, p. 19).

The presuppositions of the Santiago Table referred to the Declaration of Quebec (1984), from which the New Museology Movement (MINOM) was created, whose proposal reaffirms the social function of museums as the basis of a museological activism aimed at human development:

> While preserving the material fruits of past civilizations, and protecting those who witness today's aspirations and technology, the new museology - ecomuseology, community museology and all other forms of active museology - is primarily interested in the development of populations, reflecting the driving principles of their evolution while associating them with future projects. (Minom 2010)

Fact is that from this movement, the museums would never be the same as before. From the '90 s on, this museological movement evolved, giving way to Sociomuseology, constituted as a disciplinary area of knowledge determined to extend the traditional guidelines of Museology and to adapt the new museum to the characteristics and needs of contemporary society. The social action becomes understood as the museum's active contribution to the sustainable development of humanity based on equal opportunities and social, cultural, and economic inclusion.

The museological heritage is then understood as a sustainable development resource. In fact, for Hugues de Varine, two "provisional definitions" identify the premises for the ecomuseum (Varine-Bohan 2012, p. 14):

- That local development is a voluntary process of mastering cultural, social and economic change, rooted in an experienced heritage, nourishing itself from it and generating heritage.

- That heritage (natural and cultural, living or consecrated) is a local resource that has no other reason to be than its integration into the dynamics of development. Inherited, transformed, produced and transmitted from generation to generation, the heritage also belongs to the future.

Based on these assumptions, Varine identifies two inevitable conclusions (Varine-Bohan 2012, p. 14):

- Development will not be sustainable, and therefore real, if it is not done in accordance with the heritage, and if it does not contribute to life and its enrichment;
- Development cannot happen without the effective, active and conscious participation of the community that owns this heritage.

In order to panoramically aquilate the inflections resulting from this change in positioning that would bring about profound transformations in the way museums operate in general and not only the museums conceived from this new perspective of museal relations, the following synoptic table should be used:

CLASSICAL MUSEOLOGY	NEW MUSEOLOGY
Building	Territory
Collection of objetcs	Cultural and Natural Heritage
Scientific knowledge and interdisciplinarity	Traditional knowledge combined with scientific knowledge / transdisciplinarity
Public (scholarly and passionate)	Local residents and visitors
Knowledge production, education and entertainment	Collective knowledge building / Creative economy initiatives

Source author/Varine

One can thus conclude that of the characteristics common to all the possible typological modeling that emerged with the New Museology/Sociomuseology/Social Museology, it is noteworthy:

1. The active, creative, and collaborative participation of the population involved;
2. Actions and processes inspired by local specificities;
3. The importance of the idea of territory (lived space) as a field of action of the museum;
4. The collective appropriation of assets/collection.

Among the main lines of action of the ecomuseums and museums of territory are the protection of the natural and cultural heritage, assured singularly by the appropriation of the populations to whom it concerns its safeguard, the integral management of the territorial space and the pedagogical action of sensitization through the Environmental Education, including in this concept the Patrimonial Education, understood as "the construction of a relationship with the communities and the places, making possible the social appropriation of knowledge of which the patrimony is support" (Scifoni 2012, p. 37). In this sense, the educational actions assume a diversified and participatory character that enables the insertion of the local population in the challenge of thinking about the protection of the referential assets of their collective memory in order to provide the awareness of citizens as subjects of their own history, as the Freirean thought preaches.

There are no canonical models for ecomuseums, community museums, and all other modalities that can be understood as territory museums or in a broad and generic conception as open-air museums, but principles to be followed. In everyday practice, they differ in the way they correspond to the needs of articulation between heritage–population–territory, management methods, levels of institutionalization, and, finally, the degree of mobilization in relation to the identity processes of spatial–temporal appropriation of heritage territories.

Thus, these museums, conceived as memory management systems and agents of transformation, are responsible for the mediation between heritage (material and immaterial) and its public, between the landscape and society, based on a basic criterion of mediation that is established with respect to diversities—here covering populations, generations, languages, origins, beliefs, and experiences, always with the aim of inclusion and accessibility. In this area, several subtypologies are included. Varine then proposes to group open-air museums from their central purposes:

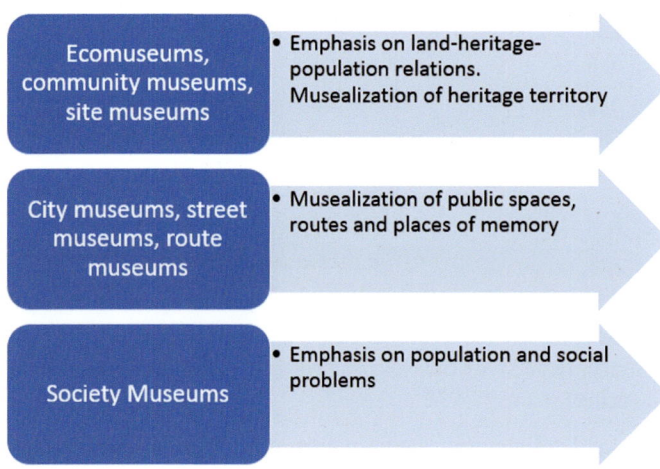

Source Varine

According to the profile of their members and the nature and specificities of their purposes, museums will shape their management and governance systems. It is desirable that they draw up a participatory inventory, with or without the support of specialists, in order to "describe, classify and define what discerns and affects them as heritage, in a dialogical construction of knowledge about their cultural heritage" (Iphan 2016). It will be necessary to make use of basic techniques of documental survey, systematization, and interpretation of data and dissemination of information. It is important to develop planning processes based on museological guidelines and, ideally, to build a collective management plan. For this, it is essential that councils are created with deliberative character and that deliberations are taken democratically. But, above all, it is essential to avoid the fetishization of objects, buildings of more goods to be preserved, through which problematizing and critical perspectives are eliminated, promoting the "reification that removes these objects from their context, the social fabric, isolating them from the set of social processes" (Scifoni 2012, p. 34).

The following table represents an attempt to identify, over time, the evolution of the main characteristics of open-air museums, even if these characteristics sometimes coexist and are articulated contemporaneously in many memory preservation initiatives:

Classification Table for Open-Air Museums

Management model	Subtypes
1st generation	
Centralized management by the body responsible for creating use and protection standards	Natural parks, society museums, site museums (archaeological, paleontological, geological and historical), zoos, botanical gardens
2st generation	
Decentralized and participative management	Ecomuseums, neighborhood museums, city museums, rural museums, route museums, street museums
3st generation	
Community self-management / networking	Community museums, favela museums, resistance museums (indigenous, Afro-Brazilians / quilombolas, riverside / caiçaras), territory museums

Source author

Finally, safeguarding the plural character of the territory and the multiple possibilities of its musealization, it can be said that the practices of Social Museology focus on the relationship of society with a territory, with a view to identity preservation, concerning the physical–geographical, environmental, and historical–cultural aspects identified by a certain social group, which proposes to participate directly and actively in the management of its cultural and natural heritage in favor of local and sustainable development.

It is necessary, however, to revisit some basic concepts. Heritage, territory, and museum are terms that are so multifaceted that we run the risk of mismanaging them. It is necessary, therefore, to make explicit the meaning with which each one of them will be considered and to what extent they will be correlated to compose the proposal of musealization of the territory.

José Reginaldo Santos Gonçalves proposes the appropriation of the concept of heritage as a category of thought, an original contribution of the anthropological tradition.

After all, human beings use their symbols primarily to 'act' and not only to 'communicate'. Patrimony is used not only to symbolize, represent or communicate: it is good to act. It mediates sensitively between human beings and divinities, between the dead and the living, past and present, between heaven and earth, among other oppositions. It does not exist only to represent abstract ideas and values and to be contemplated. It, in a way, builds, forms people (Gonçalves 2007).

In that sense, he is not neutral. The recognition of cultural heritage value has a political dimension. Cultural value is not in things. Unlike fetishization, which cultivates the object as if its value were intrinsic to it, for Gonçalves, the materiality of objects cannot be dissociated from their social and symbolic uses. "What we call cultural goods do not have their own identity, but the identity that social groups impose on them," Professor Ulpiano Bezerra de Menezes points out (Gonçalves 2007).

Values are arbitrarily assigned from proclaimed qualities. Therefore, they are not permanent, invariable, but produced in the concrete game of social relations. "There is no heritage that is not at the same time condition and effect of certain modalities of individual or collective self-conscience", insists Gonçalves.

It is common to attribute the survival of the patrimony to its pseudo-neutrality, as explained below:

> When one thinks of heritage, immediately comes to light some concepts that have crystal-lized over time, although as historical production constantly reworked, linked to the idea of monumentality, exceptionality, seniority, social heritage of a past common to all, thus eliminating all conflicts and contradictions. Thus, imbued with the idea of neutrality, it is now seen as involuntary survival of this past (Aurelio and Scalabrini 2004).

It is worth opening a parenthesis here to transpose the concept of pseudo-neutrality of heritage into the context of urban transformations that are often considered anodyne and inevitable in the face of the dynamics of progress and modernization:

> In a certain way, the decharacterization of urban environments occurs less in function of the passage of time that impose new dynamics, than of some urban projects that, since the 19th century, have transfigured cities in the name of efficiency, rationality of space, finally, in the name of the capital, displacing residents, breaking sociabilities and bonds of belonging. On the other hand, we cannot forget either that these changes that cities go through over time, adding to them the marks of each epoch, are not exactly spontaneous, especially when urban space is transformed into merchandise, as happens in capitalist societies, and real estate speculation gives the tone to urban development (Aurelio and Scalabrini 2004).

There is, therefore, no way to conceive of a proposal for neutral, apolitical terri-torial musealization. Professor Ulpiano is incisive on this point: "I find more and more grounds to believe that the museum should be the place of questions, much more than of answers" (Meneses 2011). The information society, according to him, defines hierarchies and control systems. Hyperinformation causes disinformation. The museum is and always has been an ideological support of power. Hence the option of museums of resistance and ecomuseums to adopt self-management.

In the end, in order to seek the intelligence of how the museum of territory assumes the heritage as a development resource, it is necessary to establish what is meant by resource and development. For Raffestin, "a resource is not a thing", the matter itself, it "is a relationship whose conquest brings out properties necessary to satisfy needs"

(Haesbarth 2004, p. 3). On the other hand, the understanding of what development is is based on the theoretical assumptions of sustainable social development advocated since the Stockholm Conference in 1972, then revalidated by the Earth Summit in Rio de Janeiro in 1992, Rio + 20, and the conferences that followed at international, national, and local levels (in the framework, for example, of Local Agendas 21 and Agenda 2030) that gave them legitimacy by integrating the heritage, initially natural, then cultural, in the so-called "sustainable" development plans. Basically, all these international conventions take up the principle enshrined in the document Our Common Future, Brundtland (1987), according to which sustainable development is that which aims to "satisfy the needs of the present without compromising the ability of future generations to satisfy their own needs".

In this tuning fork, the culture becomes understood in three dimensions:

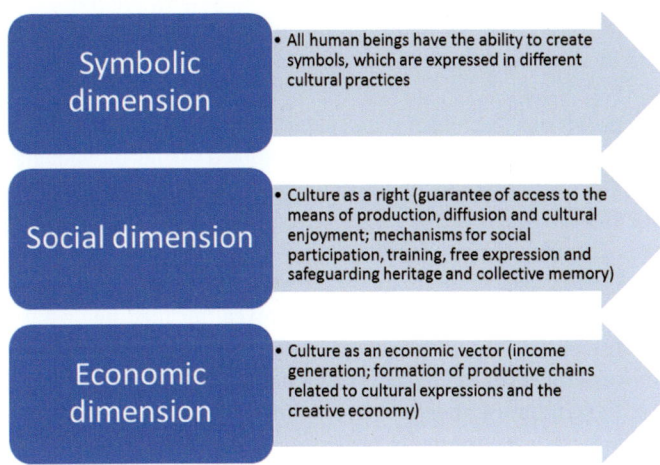

Symbolic dimension	• All human beings have the ability to create symbols, which are expressed in different cultural practices
Social dimension	• Culture as a right (guarantee of access to the means of production, diffusion and cultural enjoyment; mechanisms for social participation, training, free expression and safeguarding heritage and collective memory)
Economic dimension	• Culture as an economic vector (income generation; formation of productive chains related to cultural expressions and the creative economy)

Source II National Conference on Culture/2010

Canclini warns, however, that the relationships that define the use of these capitals are complex. By considering that the appropriation of economic capital—here identified as natural resources (land, fruits, minerals, etc.)—is the primary objective of the occupation of a territory, which implies also appropriating its people or at least the product of its workforce, Canclini rehearses the conception of a social theory of heritage:

> If we consider the uses of heritage from studies on cultural reproduction and social inequality, we see that the goods gathered in history by each society do not really belong to everyone, even if they formally appear to belong to everyone and are available for everyone to use. (Canclini 2011, p. 194)

Consequently, the formation and permanent reformulation of cultural capital is the result of a social process that, as Canclini warns, "like other capital, accumulates, restructures, produces income and is unevenly appropriated by various sectors". In this sense, the Argentine thinker adds:

> Cultural heritage functions as a resource to reproduce the differences between social groups and the hegemony of those who gain preferential access to the production and distribution of goods. (Canclini 2011, p. 195)

Caclini calls attention to the fact that the unequal character of the cultural heritage arises from the inequality that begins in the election and sacralization of the heritage, since there is no equal participation in its construction:

> Cultural heritage thus serves as a resource to produce the differences between social groups and the hegemony of those who enjoy preferential access to the production and distribution of goods. (Canclini 1994, p. 97)

Indeed, in modern societies, the hegemonic classes consecrate themselves as superiors by appropriating themselves privileged to the common patrimony. They make use of historically accumulated knowledge, count on material, economic, and intellectual resources, and thus improve their cultural products and impose their symbolic, aesthetic, and ideological values, associating them with what is identified as modernity.

Meanwhile, Canclini points out, "in recent years, demographic expansion, uncontrolled urbanization and ecological depredation have prompted social movements concerned with restoring neighbourhoods and buildings, or maintaining a livable urban space". (Canclini 1994, p. 102)

Within this spectrum of diversification, a new category of cultural heritage within this context: the cultural landscape. "Landscape is memory and palimpsest", says Michel de Certeau, to explain that landscapes allow a revealing reading, whose signs constitute direct references to the elucidative historical path through which its population has related over time with its environmental heritage.

The consideration of the Cultural Landscape as a cultural heritage and therefore as an object of protection, although originally present in tradition in geographical and other scientific discussions, nevertheless began to take shape with the 1972 UNESCO World Heritage Convention. However, it is observed, mainly through the adoption of the category of "cultural landscape" as a possible heritage, that the decision to consider in a more global way the relations between culture and space will only be consolidated in Eco Rio 92, when UNESCO establishes its concept as the interaction between the natural environment and human actions, recognized by social and cultural expressions, resulting from a territorial identity woven by a certain social group, even if this identity is plural.

Object of the right to memory and territorial identity, the cultural landscape becomes recognized as a cultural right. Indeed, in the International Charter on Cultural Tourism adopted by ICOMOS in 1999, the right to participate in the management of the territory and the landscape is recognized as an objective right:

> A fundamental objective of the territory's heritage management is to communicate its meaning and the need for its conservation to both the host community and visitors. Physical, intellectual and/or emotional, sensible and well-managed access to heritage assets, as well as access to cultural development, constitute both a right and a privilegie (Icomos 1999).

After the inclusion of the cultural landscape as a heritage eligible for UNESCO, IPHAN regulated it as an instrument for the preservation of Brazilian cultural heritage in 2009 and, on July 6, 2012, recognized Rio de Janeiro as the first urban area in the world to receive the stamp of cultural landscape. IPHAN's ordinance 127 conceptualizes the Brazilian Cultural Landscape as "a peculiar portion of the national territory, representative of the process of interaction of man with the natural environment, to which life and human science have imprinted marks or attributed values", considering "the dynamic character of culture and human action on the portions of the territory to which it applies", coexisting "with the transformations inherent to sustainable economic and social development", and valuing "the motivation responsible for preserving the heritage".

Thus, from the perspective of Brazilian legislation, "the cultural landscape is made up of the sertanejo and the Caatinga, the candango and the Cerrado, the Pantanal and the boiadeiro, the gaucho and the pampas, the fisherman and the traditional boats, the traditions of the forest and the indigenous tribes".[5]

Even today considered as an innovative way of conceiving the protection and management of the cultural heritage, making the preservation of material and immaterial assets compatible at a single time, some assets that were already listed before the regulations are recognized as cultural landscapes. This is the case of the Complexo de Áreas Protegidas do Pantanal Pantanal Protected Areas Complex (MT/MS) (Pantanal Protected Areas Complex), the Complexo de Conservação da Amazônia Central (AM) (Central Amazon Conservation Complex), Ilhas Atlânticas: Fernando de Noronha e Atol das Rocas the Atlantic Islands: Fernando de Noronha and Atol das Rocas (PE/RN) (Atlantic Islands: Fernando de Noronha and Atol das Rocas), Parque Nacional Serra da Capivara (PI) (the Serra da Capivara National Park), Parque Nacional do Iguaçu (PR) (Iguaçu National Park), Reservas da Mata Atlântica (PR/SP) (Atlantic Forest Reserves), and the Reservas do Cerrado: Parques Nacionais da Chapada dos Veadeiros e das Emas (GO) (Cerrado Reserves: Chapada dos Veadeiros and Emas National Parks).

In particular, the protection of the Atlantic Forest reserves in São Paulo territory by the CONDEPHAAT in 1985 and later inscribed by UNESCO in the list of Universal Patrimony of Humanity in 1999 represents a landmark in the speeches about the traditional history and memory of São Paulo, as Crispim records:

> The new understanding presented the landscape as a way of linking the use of the legal figure of the protection to the imperatives of territorial planning, in an operation that associated the territory with the field of cultural heritage (Crispim 2016).

It is necessary to point out that the preservationist practices prescribed for the cultural landscape are different from those of the first conservationists who intended to recreate and reinterpret the myth of earthly paradise by creating uninhabited national parks, where man could only contemplate the beauties of Nature, as advocated by the thought of John Muir, whose preservationist vision is at the root of the model of Conservation Units that has spread throughout the world, including Brazil.

> In theoretical terms in the United States in the 19th century, there were two visions of conservation of the "natural world" which were synthesized in the proposals of Gifford

---

[5]Cultural Landscape/Iphan. Accessed at http://portal.iphan.gov.br/pagina/detalhes/899/

Pinchot and John Muir. These ideas had great importance in conservationism inside and
outside the United States. (Diegues 2000, p. 28)

The concept of cultural landscape dialogues with the thought of Pinchot, a forest
engineer trained in Germany who created the movement for the conservation of
resources, proclaiming its rational use, whose ideas were precursors of what is now
called "sustainable development". This is what Nascimento and Scifoni defend more
than a century later:

When it comes to protecting the landscape, these transformations must be compatible with
sustainable forms of social and economic development in order to ensure the preservation
and enhancement of cultural heritage. (Scifoni 2010)

Obviously, the seal of the cultural landscape alone does not guarantee "the effectiveness in protecting this cultural heritage, taking into account its size as a cut of the
territory". In order to make preservationist practices effective, it will be indispensable
to establish a pact between the public authorities and society, which presupposes a
process of involvement and interlocution with the local communities that transforms
them into allies and active subjects in the consolidation of a management plan capable
of ensuring the preservation of the natural heritage, including the fauna, flora and
sustainability of the territory's inhabitants.

If the creation of the cultural landscape category broadens the scope of appropriation of cultural heritage, the transformations in the scope of museums accompany
this dynamic. And the case of the definition of museums approved by ICOM at the
22nd General Assembly, Vienna, on 24 August 2007:

*"The museum is a permanent non-profit institution at the service of society and its develop-
ment, open to the public, which acquires, preserves, investigates, communicates and exhibits
the material and immaterial heritage of humanity and its surroundings for the purposes of
education, study and delight".*

In Brazil, the Statute of Museums (Law 11,904/2009) basically adopts this same
concept, in its Article 1:

For the purposes of this Law, museums are considered to be non-profit institutions that
conserve, research, communicate, interpret and exhibit, for preservation, study, research,
education, contemplation and tourism purposes, sets and collections of historical, artistic,
scientific, technical or of any other cultural nature, open to the public, at the service of society
and its development.

Adding, however, in the Single Paragraph:

In this Law, the institutions and museological processes aimed at working with the cultural
heritage and the territory aiming at cultural and socioeconomic development and community
participation will be framed.

Conceptual enlargements, it should be noted, invariably advance in the sense of
empowering the protagonism of the museal subjects who act in the musealization
processes. "They are processes of patrimonial treatment that start from the idea of

territory, of a community living in this territory, building its patrimony and experiencing what already exists and, above all, deciding on the criteria of musealization and preservation", evaluates teacher Cristina Bruno (2000).

The expansion of the scope of museums and the broadening of musealization horizons are not exempt from conflict, being the target of intense disputes both in the academic field and within ICOM. An example of this is that at the 25th General Conference in Kyoto in 2019, ICOM was unable to put the proposal to update the concept of the museum to a vote, given the resistance of the majority of its members. For several reasons, in fact, no consensus was reached to approve the following proposal:

> Museums are democratizing, inclusive and polyphonic spaces, oriented towards critical dialogue about the past and the future. Recognizing and dealing with the conflicts and challenges of the present, they hold, on behalf of society, the custody of artifacts and specimens, by which they preserve diverse memories for future generations, ensuring equal rights and access to heritage for all people.

From these reflections, we conclude that there is not a single way to promote cultural mediation within museums by taking into account that memory, museum, and heritage are fields of political dispute and inseparable from oblivion and power. Whatever the mode or type of museum, what matters is the relationship that has been established between heritage, territory, and the community that is the protagonist, by identifying, studying, and safeguarding the local heritage by placing it at the service of the communities and their sustained development.

Acting as a system of memory management and an agent of social transformation, all the forms of musealization mentioned here establish the exercise of mediation between heritage (material and immaterial) and its public (residents and visitors), between the landscape and society, guided by respect for diversities—here covering the different generations, knowledge and doings, origins, beliefs, and experiences, always with a view to inclusion.

From this point of view, since they are established as instruments for the social and cultural development of their communities, the museological processes constituted in the context of the appropriation of railway heritage, as well as in community museums, ecomuseums, and territory museums, are referred to by the concept defended at the Santiago de Chile Round Table, according to which heritage must be considered in its entirety, especially with regard to the urban context in which they are inserted. In this sense, they are experimenting with new paradigms of action that demand the involvement and support of society and public authorities through museological processes, whose understanding and analysis still require inter and multidisciplinary investigations.

It is possible to conclude, however, that the social appropriation of the railway heritage, whatever the mode of preservation to be adopted, will inexorably have to dialogue with the city and its social fabric, considering its multiple territorialities. Besides dialoguing with the most pressing issues of citizenship, it is necessary to avoid the fixation on the commonplace of the appreciation of the nostalgic feelings of a gloriously idealized past. In addition to interacting with citizens, focusing on the relationships of individuals with each other and with the space they produce and its meanings, it is necessary to reflect on the forms of appropriation of this past with a view to building bridges with the present and the future.

References

Aurelio CR, Scalabrini MV (2004) Patrimony and city. "Survival" of the past in Ribeirão Pires. Arquitextos, São Paulo, year 04, n. 048.07, Vitruvius. https://www.vitruvius.com.br/revistas/read/arquitextos/04.048/587

Barbosa F (2015) Human rights, cultural heritage and public policies. In: Bens cultural and human rights/ org. Inês Virgínia Prado Soares and Sandra Cureu. Edições Sesc São Paulo, São Paulo

Birth FB, Scifoni S (2010) The cultural landscape as a new paradigm for protection: the experience of the Ribeira Valley-SP. CPC Magazine, São Paulo, no 10, pp 29–48, May/Oct 2010. http://www.revistas.usp.br/cpc/article/view/15660

Bolaño C et al (2012) Incentive laws for culture through fiscal resignation in Brazil. In: Cultural policies: research and training/ organization of Lia Calabre. - Itaú Cultural; Rio de Janeiro: Fundação Casa de Rui Barbosa, São Paulo

Brazil (2020) Constitution of the Federative Republic, 1988. Official Diary of the Union, Brasília, DF. http://www.planalto.gov.br/ccivil_03/Decreto-Lei/Del0025.htm. Accessed 03 Aug 2020

Bruno MCO (2000) Museology: the struggle for the persecution of abandonment. São Paulo, 2000. Tesis de Livre Docência do MAE-University of São Paulo, 238p

Bruno MCO, Neves (Orgs.) KF (2008) Museums as agents of social change and development—museological proposals and reflections. Xingó Museum of Archaeology, Saint Christopher, p 210

Canclini NG (1994) The cultural heritage and the imaginary construction of the national. National Historical and Artistic Heritage Magazine. Brasília: Institute of National Historical and Artistic Heritage, no 23, pp 95–115

Canclini NG (2011) Hybrid cultures—strategies for entering and leaving modernity. Translation by Ana Regina Lessa and Heloísa Pezza Cintrão. São Paulo: EDUSP, 4th and 5th reimp

Certeau M (2013) The invention of everyday life: 1. Arts of doing; 20th ed (transl by Efraim Ferreira Alves). Voices, Petrópolis, RJ

Chagas MS (1994) Novos caminhos da museologia - Sociomuseology Notebooks no. 2. Lusophone University of Humanities and Technologies, Lisbon

Chagas MS (1999) There is a drop of blood in each museum: the museological viewpoint of Mário de Andrade. Sociomuseology Notebooks, no. 13. Lusophone University of Humanities and Technologies, Lisbon

Choay F (2006) The allegory of heritage (transl by Luciano Vieira Machado), 4th edn. Liberdade Station: UNESP, São Paulo, 288 p

Crispim FB (2016) Between geography and heritage: study of the actions of preservation of the São Paulo landscapes by the Condephaat (1969–1989). São Bernardo do Campo. UFABC/Fapesp Publishing House, pp 138–139

Diegues ACS (2000) The modern myth of untouched nature – 3th edn. Hucitec Support Center for Research on Human Populations and Brazilian Wetlands, USP, São Paulo

Freire P (1987) Pedagogy of the oppressed, 17th edn. Paz e Terra, Rio de Janeiro

Ghirardello N (2002) On the edge of the line: urban formations of Northwest São Paulo. Unesp Publishing House, São Paulo

Gomes AV, Oliveira ER (2019) Acervos e Ferrovia: preservation of São Paulo's railway documental heritage: http://web.revistarestauro.com.br/acervos-e-ferrovia-preservacao-do-patrimonio-documental-ferroviario-paulista/ Accessed 06 Aug 2019

Gonçalves JRS (2007) Antropologia dos objetos: coleções, museus e patrimônios. MinC, Rio de Janeiro

Halbwachs M (2003) The collective memory (transl by Beatriz Sidou). Centauro, São Paulo

Icomos (1999) International charter on cultural tourism, Mexico City. https://www.icomos.org/en/newsletters-archives/179-articles-en-francais/ressources/charters-and-standards/162-intern ational-cultural-tourism-charter. Accessed 25 Feb 2020

Iphan (2016) heritage education: participatory inventories - application manual/Iphan, Institute of National Historical and Artistic Heritage, Brasília. http://portal.iphan.gov.br/uploads/publicacao/inventariodopatrimonio_15x21web.pdf. Accessed 25 Feb 2020

Iphan (2020) Cultural Landscape. Institute of National Historical and Artistic Heritage, Brasília. http://portal.iphan.gov.br/pagina/detalhes/899/. Accessed 25 Feb 2020

Kaseker DP (2014) Museum, territory, development: guidelines for the musealization process in the management of Itapeva (SP) heritage. Unpublished Master Thesis, (Master in Museology). Museology, Universidade de São Paulo, São Paulo. https://doi.org/10.11606/d.103.2014.tde-090 22015-115653. Accessed 25 Feb 2020

Kaseker DP (2018) Museums from São Paulo and Railroad Memory. In Latin American Heritage – Interdisciplinary Dialogues on Brazilian and Argentinian Case Studies. In: Lopes da Cunha F, Rabassa J, Santos M (eds) The Latin American Studies Book Series. Springer International Publishing. https://doi.org/10.1007/978-3-319-58448-5-1

Lanna ALD (2016) Cities and railways in 19th century Brazil - some reflections on the diversity of social meanings and urban impacts: Jundiaí and Campinas. http://unuhospedagem.com.br/rev ista/rbeur/index.php/shcu/article/viewFile/883/858. Accessed 20 Jan 2016

Magnani JGC (2018) apud CONDEPHHAT 50 Years: Records of a trajectory/ Martins AL (org.). São Paulo: Museum of Sacred Art of São Paulo. p. 92

MARTINS (org.). AL (2018) CONDEPHAAT 50 Years: records of a trajectory. Museum of Sacred Art of São Paulo, São Paulo

Mazzoco (2005) apud CRUZ, Thaís F. dos Santos - Paranapiacaba - The architecture and urbanism of a railway village

Meneses UTB (1994) From the theater of memory to the laboratory of history: the museological exhibition and historical knowledge. Annals of the Paulista Museum. New Series, vol 2, pp 9–42, Jan/Dec

Meneses UTB (2011) Communication/information at the museum: a review of premises. I Seminar Information Services in Museums. São Paulo, Pinacoteca do Estado

Moutinho M (1993) About the concept of Social Museology. Sociomuseology Notebooks, no. 01. Lisbon: ISMAG/UHLT

Minom M (2010) Declaration of Quebec – Basic Principles of a New Museology 1984. Cadernos de Sociomuseologia, 38

Perez R (2019) Inventory of Steam Locomotives in Brazil - Railway Memory. http://www.memori aferroviaria.com.br/ Accessed 08 June 2019

Scifoni S (2012) Education and cultural heritage: reflections on the theme. In: Educação patrimonial: reflexões e práticas./ Átila Bezerra Tolentino (Org.) Superintendence of IPHAN of the Paraíba, João Pessoa

UNESCO (1948) Universal Declaration of Human Rights. New York, 1948. https://unesdoc.une sco.org/ark:/48223/pf0000139423

Varine-Bohan H (2010) About the Round Table of Santiago de Chile (1972). In: O ICOM-Brasil e o Pensamento Museológico Brasileiro - selected documents, vol 2. State Pinacoteca: Secretariat of Culture: Brazilian Committee of the International Council of Museums, São Paulo,, pp 38–43

Varine-Bohan H (2012) The roots of the future - The heritage at the service of local development. Translated by Maria de Lourdes Parreiras Horta. Porto Alegre: Medianiz

Zagato JAC (2020) The thematic study of the railway heritage of the State of São Paulo. In CONDE-PHAAT 50 Years: records of a trajectory/ MARTINS, Ana Luiza (org.). Museum of Sacred Art of São Paulo, São Paulo, pp 149–154

Zambello MH (2005) Railroad and memory: study on the work and category of the old railroad workers in the Industrial Village of Campinas. Supervisor Dr. Heloísa Helena Teixeira de Souza Martins. Unpublished Master's Thesis, University of São Paulo

Sorocabana Station in Bauru (São Paulo, Brazil): Diagnosis for Restoration

Nilson Ghirardello

Abstract Bauru is a city located in the State of São Paulo, southeastern Brazil; created at the end of the nineteenth century as part of the territorial occupation process of western São Paulo, where coffee plantations for export, followed by the railroad, played a key role. Three railroad companies were installed in the city: Companhia Sorocabana (1905), Northwest of Brazil (NOB) (1905), and Companhia Paulista (1910). The most expressive for the urban development of the city was the Northwest of Brazil (NOB) railway company, due to its headquarters in Bauru, its size, and the significant strategic and pioneering role toward the west of the country. However, the NOB company only established its headquarters in the city due to the pre-existence of the Sorocabana Company station. The historical importance of this small construction, still existing, is its size; in 2015 a project for its restoration was proposed. There are two main objectives of this work: to draw a history about the installation of the railroads in Bauru and to present the diagnosis of conservation of the building of the old Sorocabana station, used for the elaboration of the restoration project, but unfortunately, not yet executed.

Keywords Brazil · Bauru · Railroads · Station · Restoration

1 The Initial Religious Patrimony and the Urban Layout of Bauru

As part of the coffee expansion, quickly followed by the railroads in São Paulo, a state in the southeast of Brazil, the city of Bauru was created. Like so many others in the interior of this state, it was originated, during the nineteenth century, from the area of land donated to the Catholic Church, by a great farmer, with the purpose of urban formation, in the year 1884. This process, where the urban lands are under the

N. Ghirardello (✉)
Faculty of Architecture, Arts and Communication, State University "Julio de Mesquita Filho"—UNESP, Bauru, Sao Paulo, Brazil
e-mail: nilson.ghirardello@unesp.br

© The Author(s), under exclusive license to Springer Nature Switzerland AG 2021   167
F. Lopes da Cunha and J. Rabassa (eds.), *Festivals and Heritage in Latin America*,
The Latin American Studies Book Series,
https://doi.org/10.1007/978-3-030-67985-9_10

possession of the Church, which gave them to those interested for use and enjoyment and not full ownership, received the name of Religious Patrimony (Ghirardello 2010).

The first buildings of the village were built next to a road that linked the closed forests of the west of the state to two neighboring urban centers, which had been constituted previously. This road, which became a street called Araújo Leite, was located in a north-south direction, near the "Das Flores" river, in one of the limits of the village.

The figure of the area where the village was formed resembled a triangle, on a slope, whose boundaries were limited by two watercourses to the north: the Bauru River and its affluent, the "Das Flores" Creek. Closing the area, to the south, there was a great straight line where later a commercial route was formed. In all this area, the reticulated layout of the city was defined, formed by regular streets and blocks. In 1893, another area of land was annexed to the first, from a new donation to the Catholic Church by another farmer, and the street was expanded, continuously, toward the south.

On August 1, 1896, the village became the seat of a municipality, the last at this point in the state and due to this, it had the astonishing area of 1,936,000 hectares (19.3 sq. km.), heading west, reaching the banks of the Paraná River, on the border with the state of Mato Grosso do Sul (Pinheiro 1928). This entire area, where there was the rarefied presence of white men and a great indigenous presence, gradually destroyed, would only be occupied at the beginning of the twentieth century, for the construction of the railroads.

Little by little, the city's population grew and, at the same time, new farmers started to occupy the region's lands, aiming at coffee farming, guaranteed by its future valorization, because the trails, drains of production, foresaw its arrival in the younger city.

At the end of the ninteenth century, the government of the State of São Paulo issued two decrees that granted the railway companies, the Paulista Company and the União Sorocaba and Ytuana Company permission to build railway lines to Bauru. In the same period, another railroad was being planned by the Federal Government that would link the State of São Paulo to the State of Mato Grosso. However, the proposals on the possible starting points of the new railroad were still divergent.

Although the coming of the railroad companies, Sorocabana and Paulista, was already assured by state decrees giving area privileges to these companies, their arrival in Bauru was still uncertain, depending on economic and political factors. The favorable signs were the increase in the number of coffee plantations in the new municipality and rumors, even if remote, of a railroad that could start in Bauru, bound for the state of Mato Grosso, to the west, and from there, through neighboring countries, it could reach the Pacific Ocean (Fig. 1).

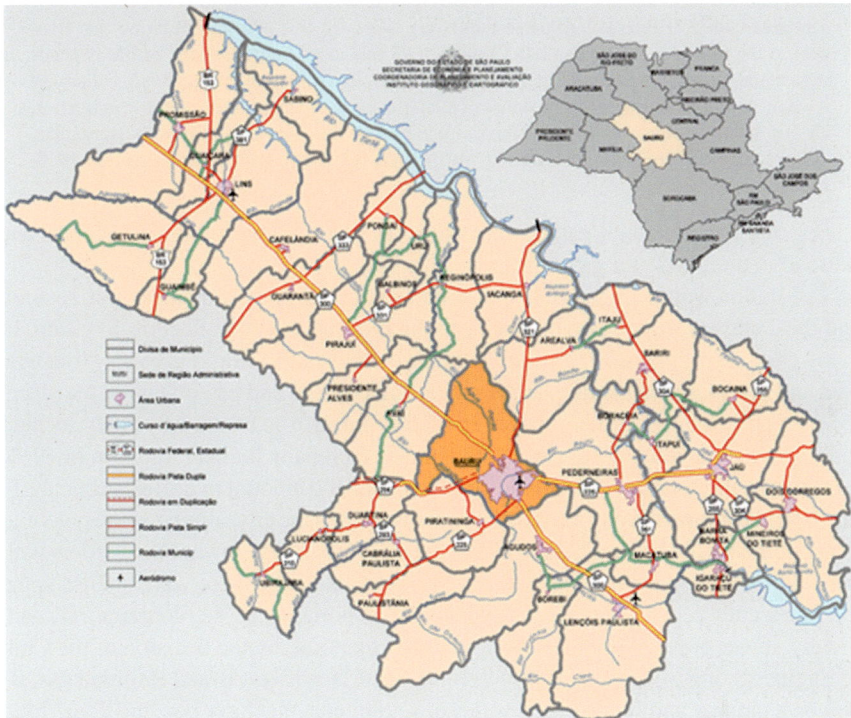

Fig. 1 Larger image, current boundaries of the Municipality of Bauru, with the urban area in the center. Above, the Municipality registered in the State of São Paulo. *Source* http://www.sp-turismo. com/mapas/bauru.htm

2   A New Moment, the Arrival of the Railroad in Bauru

At that moment, the tracks were as relevant for the communities as for the crops. And it can be affirmed, with a greater degree of comprehensiveness, because they were responsible for the development of these villages and small towns, which from them, it could reach other stages of growth.

Matos (1990) is one of those authors that referred to the value of the railroad to the cities:

> The arrival of the trails is almost always a milestone in the history of a city. With the railroad, it comes all the equipment it requires, especially when the city, for some reason, is chosen to host any special activity on the road: warehouse, workshops, offices, train crossing point or place change of trains (Matos 1990, p. 197).

In the Illustrated Album of the Paulista Company (Pérez 1918), published in such year, the relationship between the development of the State and of the entire urban network with the Company is explicit, and by extension, the State's debt to the railway companies:

The area called "Paulista", all these numerous cities, all this active population that lives next to the railroads of the Paulista Company, all this in all their relations of life depends substantially upon the prompt, regular and fast circulation offered by the railroad. Otherwise it would not be conceivable the existence of all those crops, all those cities, all those industries, all that active population. Take off the railroad and all this will disappear and disorganize, disappearing the entire population that was expanding along with its trails (Pérez 1918, Introduction).

For this reason, small estates and cities, as in the case of Bauru, received the railroad in their soil as a gift of life.

The station would obey the classifications of the railroad companies: first, second, and third classes, which corresponded to the movement and collection foreseen. In many cases, with the passing of time, the last class station would give way to a new one, and first class, depending on the movement of loads and passengers of the place.

The Sorocabana Company arrived in 1905, coming from two nearby cities, located to the east, Botucatu and Lençóis. The Paulista Company builds its station in 1910. At the time, the Bauru region already produced a good amount of coffee, particularly in the farms around the city, but the railroads moved to the city, aiming at the potential of the immense area of the State still formed by virgin forest, toward the borders with the State of Mato Grosso. For the occupation of its lands, there were studies since the nineteenth century that demonstrated the importance of a colonizing railroad. Finally, it was created in 1905, having its starting point, by the decision of the Club of Engineers of Rio de Janeiro, also in Bauru. The Northwest Brazil Railroad (NOB) will be a pioneer and will create a series of settlements that have become flourishing cities, born at the edge of the line and its stations (Ghirardello 2002).

Bauru in a short time becomes an important junction, entrance of a great region, served by the NOB, the only one to transport immense quantities of coffee that started to be produced from the banks of the Paraná River, displaced to Bauru and from there transported by the Sorocabana or Paulista Companies. The meeting of several railroads puts the city in direct contact with several regions, a privilege only compared to the state capital, São Paulo. This privileged situation transformed Bauru into a regional pole, of a considerable and central part of the State, favoring the commerce, the rendering of services, and, consequently, its growth.

2.1   The Implantation of Sorocabana Station in Bauru

Sorocabana marked its station in Bauru in August 1904 and inaugurated the line in 1905. In the Company Report, of the year 1905, there is the following excerpt about it:

During 1905, the extension of the 11,087 km line was increased, extending the stretch between Conceição and Baurú stations on the Tronco line, delivered to the provisional traffic on April 22, and to the definitive one on July 1, 1905. At this point began the Northwest Brazil Railroad (NOB), a federal concession, which has as its guideline Tieté, while in the territory of São Paulo, and by final objective Cuyabá, capital of the State of Matto-Grosso (http://www.estacoesferroviarias.com.br/b/bauru-efs.htm).

It can be observed, therefore, that NOB, started its works in the city, right after the coming of the Sorocabana Company, and the construction of its station and other support facilities were executed next to the Sorocabana station, forming already from this moment, a railroad set.

The Sorocabana line entered the south of the city, parallel to the Água da Ressaca stream, continuing to the Ribeirão Bauru, taking advantage of the flat areas of the lowered ones, in order to avoid ramps that would burden its construction cost. The route also required fewer expropriations, and the flooded land would have a low price. Construction in this sector was few because the densest occupation was in the region of Araújo Leite Street, the other end of the small town.

With the right of expropriation assured in a contract with the State, the railroads passed through where they were best suited, regardless of the small towns or cities, which due to the importance of the railroad, accepted the proposed route unconditionally. This is clear from the only formal request for the transfer of land, which arrives a few days before its inauguration, after the civil works were already executed.

"Approaching the moment of the arrival of the trails of this road to this locality, I come to ask Your Lordship to cede the land of the patrimony, necessary for the installation of the station and dependencies, figured in the attached plant, limited on one side by the blue line, MN, of 770 m of extension, that cuts through the center the blocks 138, 141, 144, 147 and, on the other side by the stream".

"Between the blue line MN and the red parallel that limits the patio of the Station, there will be an avenue along this patio, with great benefit for this locality. The road will take you to indemnify the private properties already existing on the land that you need, but you expect that this municipality will give you free assignment of your rights over the land (…)" (Letter sent to the City Hall by the Sorocabana Railway - February 27, 1905, Bauru City Hall Archive. Bauru 1905).

Unfortunately, the mentioned map does not exist anymore; however, it can be noticed that the railroad defined the area at its discretion, even establishing a new avenue in front of its station, now Pedro de Toledo Avenue and significantly named at the time, as Sorocabana Avenue.

As the railroad tracks were installed next to the Bauru River, in areas difficult to be used for housing, since they were flooded, they reinforced the segregation guaranteed by the natural relief in the shape of a valley. The future appearance of neighborhoods to the north and west, they would have as barriers of difficult transposition the rivers and patios, isolating them in a harmful way from the central area of the city (Fig. 2).

Sorocabana was fundamental for the urban development of the city; because of it, NOB launched its trails from Bauru, at the end of 1905, under the indication of the Rio de Janeiro Engineering Club, which designated that its beginning should take place in the last station docked in a city west of São Paulo State, in the Bauru case.

In 1939, with the inauguration of the new and large NOB station in Bauru, Sorocabana and Paulista stations were transferred to the building, a rare fact in Brazil, where three railway companies started to share the same space. With this, the old stations are now used for administrative services in the case of the Paulista Company, and deposits in the case of the Sorocabana Company. This fact saved them from drastic reforms or even demolitions.

Fig. 2 Picture that shows images of the city of Bauru, in 1918. As a whole we can see, among other pictures, the stations of Paulista Company, Sorocabana Company, and Northwest Brazil (NOB). (Pérez, 1918)

In 1957, the Rede Ferroviária Federal S.A. (RFFSA), a mixed company formed to concentrate the Union's railway assets, took over the old NOB. In the following years, the railroad, as well as the entire national railroad system, experienced its slow agony due to competition with automobiles, disastrous administrations, and lack of interest from the state in direct investments.

In 1962, a discussion began about the creation of a company to control the São Paulo railroad network, and gradually some of the state's railroads were absorbed by the government, until in 1971 Governor Laudo Natel created Ferrovia Paulista S/A (FEPASA), which absorbed the entire São Paulo railroad network.

In 1998, FEPASA was federalized and fully incorporated into the Rede Ferroviária Federal S.A., and the former FEPASA railroad network is now called Malha Paulista.

In the 1990s, the passenger trains were definitively extinct, and in that same decade, the old RFFSA was privatized. For the railway network that passes through the city of Bauru, at that time, the concession was taken over by the Novoeste Company, which unoccupied the central station of the old NOB and the workshops, and also the old stations of Sorocabana and Paulista, considered unnecessary for operational use. Since then, no significant work was done, and all the stations and other buildings were abandoned, and only the rails and the road are precariously maintained.

In 1998, Novoeste was merged with Ferronorte and Ferroban. In 2002, new transformations took place, and from there Novoeste do Brasil was born, until 2006, when

Novoeste do Brasil was merged, and Brasil Ferrovias with América Latina Logística (ALL), which began to manage only the transportation of cargo on the line. However, the maintenance of the railroad network, in the Bauru region, was very precarious, with derailments that forced the Federal Public Ministry to file civil actions against Novoeste/Malha Oeste and Ferroban/Malha Paulista, belonging to América Latina Logística. Later on, this control changes hands again, in relation to the west mesh, passing to Rumo Company, Cosan Limited logistics arm.

Due to the historical and architectural importance of Bauru's railroad complex and its abandonment, it was opened, in the beginning of the 1990s, a study for preservation and toppling, a term used for the integral protection of the property, by the responsible organ of the State of São Paulo, the Historical, Archaeological, Artistic and Tourist Heritage Defense Council of the State (CONDEPHAAT). The request made only in 2017 targeted the entire Bauru railway complex, built by all three companies that have established themselves in the city and consisting of stations, working villages, workshops, administrative buildings, in addition to the esplanades. (Process n° 30,367. Subject: Calls for the toppling of buildings belonging to the headquarters of the Old Railroad Northwest of Brazil, located in Bauru).

The innovative measure, in the sense of preserving a set of railroad assets and not only the station, as was previously the case, showed that the complex of buildings was relevant, as well as all the actions historically developed in those spaces, both those aimed at the use of the means of transportation by people and those oriented to the world of work, seen in all its hierarchies. With the criterion adopted it was clear that the preservation of the stations had a weight similar to that of the workers' villages, or to the workshop sheds.

During that time, the old NOB offices became the Regional Railway Museum, and years later, the large station of the same company was transferred to the City Hall, which used it, even partially, for renting cultural sectors.

Among all the goods belonging to the railroads that arrived in the city, the need to recover the old station of Sorocabana, due to its precarious state of conservation, stands out. Another factor that indicates its restoration is that the small building, of historical and architectural interest, represents, above all, the genesis of the urban development of the city, because it was, the main reason for the coming of the other two railroad companies to the city, among which the most relevant of them, the NOB, which propitiated the full growth of the city of Bauru.

Even the railway station, being in general the most architecturally elaborated building, due to its visibility and public access, it is also part of the railway complex, being therefore the result of the culture and industrial patrimony. The definition of industrial heritage, according to ICOMOS, is as follows:

(...) the traces of industrial culture that have historical, technological, social, architectural or scientific value. These traces include buildings and machinery, workshops, factories, mines and treatment and refining sites, warehouses and storage facilities, production centers, energy transmission, means of transport and all their structures, as well as places where social activities related to industry have been developed, such as housing, places of worship or education (Icomos: Letter of Nizhy Tagil 2005; Kuhl 2008 p. 51).

The industrial heritage became more valued around the 1950s in England when an inventory was made and recognized by the National Council of Archaeology. Another important moment for these assets was with the creation of the International Committee for the Conservation of Industrial Heritage (TICCIH), an organization whose objective is to

> (…) promote international cooperation in the field of preservation, conservation, location, research, documentation and enhancement of industrial heritage (Book of Abstracts of the VI Latin American Conference Ticcih 2012).

The preservation of industrial assets in Brazil dates back to the 1970s, when the fight for the preservation of railroad and industrial constructions often demolished by real estate speculation began, in search of large areas for the construction of commercial and residential complexes.

It can be said that the old railway station of Sorocabana is the first construction of industrial character of Bauru, after all if there were others before 1905, (as manufactures or ceramics) were destroyed, but certainly, the station is of a unique importance for the city. Even because small stations in large cities are rare, because they were almost always demolished to give way to bigger ones in the same place. In Bauru, the transfer of Sorocabana's facilities to the big NOB station made the small station, transformed into a deposit, miraculously preserved.

In order to start the restoration project, it was necessary to survey the physical situation of the old Sorocabana station, which we will summarize below.

2.2   General Diagnosis and Current State of Conservation

In order to make a general diagnosis of the façades and roof, we verify the state of conservation (during the year 2015) and cross it with data from documentary research, old plants, photographs, oral reports, and surveys made "in loco", in order to perceive interventions made over time.

We consider as very good the definition found in the Project Norms of the State Institute of Historical Heritage of the State of Minas Gerais (IEPHA/MG), which we will try to apply in the development of this work:

> This stage has the objective of knowing and analyzing the building under the historical, aesthetic, artistic, formal and technical aspects. It also aims to understand its current meaning and over time, to know its evolution and, especially, the values for which it was recognized as cultural heritage. The wider range of aspects will enable the perfect knowledge of the monument, indispensable to propose appropriate solutions in each case (Project Norms. IEPHA/MG 1980).

The internal spaces of the old station have been greatly modified with time, leaving little of the internal division of environments. If we consider most of the stations in this class, it should have a passage that would connect the two main façades (front and back) and that would give access to the embarkation/disembarkation of passengers, waiting, a place to buy tickets, station chief's room, and telegraphs and deposits.

The original plan of the station was not found, and even the hypotheses that we could have about its original internal division are only based on other similar constructions of Sorocabana, and on the composition of similar spaces of other railway companies, however, it was decided that it would not start for hypotheses without proof, or even for suppositions, without support in documents, or solidly supported iconography.

Externally, in the same way, we do not find the drawings of the original façades, however, we can be guided by two rare old photos, which show the general volumetry, where some details of openings and gaps are also visible. Another third photo, from the 1960s, gives us some clues of its transformation process, however, the two old images of the station were found.

Still, visualizing the existing façades, it is possible to notice marks in the apparent masonry, with which it is built, which indicate changes in the original design, and changes in the openings.

The fact that the station was the subject of few photographs can be explained due to its small size and formal simplicity. A preponderant factor, too, was its transference, from 1939 onward, to the large NOB building, which started to share the three railroads that arrived in the city.

2.3 Description and Diagnosis of External Areas

The old Sorocabana Company station is a small building of 208.64 m^2, of construction, (without the platforms coverage area), in brick masonry, of rectangular shape in the proportion of three to one, therefore an elongated rectangle, particularly if we consider only the closed areas, without the platforms. The construction of the station was made in masonry of apparent ceramic bricks with grouting in frieze mass, usual in Sorocabana's constructions in this period. The brick tying is of the "English or Gothic fit" type, where an even row and an odd row are placed, successively, and the unit size of each brick is 0.28×0.07 m.

Currently, the construction has an external mortar bar on all façades at a height of approximately 1.50 m, this element did not originally exist, as it can be seen in the old photos.

It has the main façade to the east, (Façade E1), facing Pedro de Toledo Avenue, an average distance of 31 meters, however, at a much lower level than the road, and the other façade facing the tracks to the west (Façade E3). Most of the railway stations obeyed this logic, which accompanied the design of the esplanade areas (terrains in the shape of large rectangles) and the rails themselves. As the train is a linear-shaped vehicle and needs to board and disembark passengers at the same time, the buildings in an elongated shape responded very well to this, in a clear relationship between the shape and function of the building.

It is a construction of clear industrial character, not only because of the railroad use, but also because of the materials used, simple and apparent; for the discreet ornamentation, as well as for the use of the constructive modulation. It is organized by the division into spans, which we will call masonry cloths, of equal size, between pilasters that give reinforcement to the structure and establishes a certain rhythm to the building. Next to Façades E1 and E3, we find six cloths and five pilasters, besides two wedges in the corners, respectively. In Façades E2 and E4, we find two cloths and a pilaster and two wedges in each corner, respectively. The modulation next to Façades E1 and E3 is transported to the scissors of the roof, that accompany it, as well as the French hands that are supported in the five pilasters.

The deposit, next to it, built in the 1960s and reformed by Embrazem warehouse, in the 1990s, partially connects the old Sorocabana station by Façade E4, however, without internal communication with it, nowadays. This construction, for being posterior and without any historical or architectonic value, prevents the visualization and compromise the integrity of the old station must be completely demolished.

In relation to the apparent masonry, we verify that several bricks, in the four faces, are eroded, by the action of the weather, misuse, and lack of maintenance.

The structure, in general, is in good condition, with no major cracks or cracks, having rare cracks. This is one of the positive characteristics of this and most of the railway constructions, which due to the high impact of the trepidation of this means of transportation, are generally designed to be submitted to these forces and, therefore, are extremely solid and well built.

Another subtle element and almost always little considered, but important for the whole are the access ramps to the existing platforms next to Façades E2 and E4, where the loads mainly arrived (Figs. 3, 4 and 5).

Next to Façade E4, due to the construction of the warehouse, the structure almost disappeared.

2.4 Façade E2

On this façade, facing north, we find two closing cloths between pilaster and dagger, having in one of the cloths a metal vitro and in the other a small wooden frame for energy input, both posterior. Such clues are revealed by the materials of the same, photographs of the time, and by the marks of the old windows disposed there, verified by the posterior closings of its spans that do not give continuity to the initial tying of the masonry and also by the existence, still today, of bricks seated on top of windows. On these lower cloths, a triangular pediment is placed, bordered by jagged bricks and which has a small eyepiece in its eardrum, currently fenced with bricks, located slightly above the center. Below it, when the construction served as a station, there was the identification of the city, "BAHURU", written on a wooden plate, as can be seen in photos of the station, on the plate we can find no trace of its existence today, however, we can still see in the one fixed to the bricks small metal supports that would attach to the wall, originally, the said plate. The pediment, at its apex, is topped by a rectangular element in masonry, typical of Victorian-speaking ridges, and on it is an ornamental ironwork. The whole set is surrounded by an apparent brick cornice (Fig. 6).

Fig. 3 Old photo of Sorocabana station of Bauru, without precise date. Viewed by the Façade facing the city, to the north, which we call E2, next to the tracks, observe in this one, the original openings with guillotine-type windows. It is verified that the general volumetry, at present, has changed little. *Source* http://www.estacoesferroviarias.com.br/b/bauru-efs.htm

Fig. 4 Old photo (without date) of the Sorocabana Station of Bauru, seen by the Façade facing south, which we call E4, (where it is connected to the "Embrazem" warehouse), it should be observed that on this face, unlike E2, probably there was no opening in any of the cloths. *Source* http://www. estacoesferroviarias.com.br/b/bauru-efs.htm

Fig. 5 Picture of Sorocabana station in Bauru, 1960s, seen from E2 face, facing the city. Note that the two original openings were still present in that period, however, no longer with guillotine windows but with metal windows. *Source* http://www.estacoesferroviarias.com.br/b/bauru-efs.htm

2.5 Façade E4

This façade, facing south, connects the "Embrazem" deposit to be demolished, at this point the land was grounded, in almost sixty centimeters, to reach the deposit quota. The two constructions remain independent, because the original wall of the station remains next to that of the deposit, of later execution. We have not verified the original marks of openings in this Façade, currently there is a sliding door for access, in one of the masonry cloths, certainly of posterior execution, being that the other cloth is without visibility, because, the wall of the deposit of the years 1960, covers it. Internally, in the space of the old station, there is evidence of opening, today sealed, however, also seems to us posterior.

The only old photo from this angle does not show, or is not clear from physical evidence, any kind of opening in this facade, which may show that internally it originally worked, even at the time of the station, a deposit whose use would dispense with doors or even windows to the outside environment (Fig. 7).

Fig. 6 Façade E2 (facing the city, north face), showing the brick veneers of the two original openings, and in the opening on the right was implanted the metal vitro, in a larger dimension than the original span. On the left is the unused energy box and the visible marks of the old window opening, currently sealed. Photo of the author

2.6 Façade E3

This façade, facing west, has six cloths of closing masonry, pilasters, one of which is blind, with marks of posterior intervention in the rows of the bricks, or of some type of repress cured by reform. Three cloths have two-leaf sliding doors, with sliding by internal rails, in poorly maintained wood, and the other cloths each have a metal door and a wooden door, both of which open. The three sliding doors of varied design (in panel, scale, and smooth) obey to the dimension of the brick lining on top, existing on them, the other two smaller doors are posterior and are in a bad state of conservation, and the wooden one does not complete the whole span, having on it a kind of masonry flag and on its side a masonry closing (Fig. 8).

Next to this façade is the old embarkation and disembarkation platform, elevated 0.70 cm in relation to the still existing rails and whose base is formed by apparent basalt stones with joints jumped in mortar, having its floor cemented.

Fig. 7 Façade E4, facing
south, connected to
Embrazem deposit. Note the
door implanted at the back
and the embankment made to
give access to the deposit. In
the access door to the old
station, next to Façade E4,
see cracks in the masonry
area completed at the rear,
denoted by the difference in
types of bricks. On this face,
we do not find apparent brick
cracks as in the opposite.
Photo of the author

2.7   Façade E1

As in Façade E3, this one, facing east, consists of six masonry cloths between
pilasters, one of them blind next to the deposit, three of them with two sliding doors,
in wood, in a bad state of conservation. In the two successive cloths, we find an old
guillotine window originally coming from the E2 façade, and in the next cloth, a
metal vitro and a simple wooden door to open.

It is noticeable by the dimension of the top brick lintel over the doors, that in the
cloth where the wooden window is located there was already a two-leaf sliding door,
because the lintel has the same extension of those over the other sliding doors. In the
same way, the absence of wicker in the first masonry cloth where today is the vitro
and door leads to suppose that the cloth would be blind. In other words, originally,
at least at some point in the building, there would be two blind cloths on the sides
and four with sliding doors in the center.

Next to this elevation, there is a circulation with a cemented floor similar in
dimension to the platform, however at the same level of the access, where a ramp
was built toward the Embrazem deposit attached. It should be noted that at this
junction point between the station building and Embrazem warehouse is where the
greatest amount of humidity, therefore with signs of biodeterioration, becoming clear

Fig. 8 Detail of the old boarding platform, next to Façade E3 and openings and barred, next to Façade E3. Photo of the author

by the stains and marks on the walls of the building. Also in this facade, we find several burnt bricks just above the plastered bar (Fig. 9).

2.8 Roof

The roof, in two waters, is formed by clay tiles, known in Brazil as "French" type. Currently, they are covered by microorganisms and microflora, which shows the humidity of the place and the end of its useful life. The total area covered by the station is 358.82 m^2, in its current aspect, with platforms totaling 44.91 m^2 each. These are supported on conventional wooden frames consisting of laths, third claws, and five apparent simple wooden scissors, with trusses, in a reasonable state of conservation. The roof tiles are in a bad state of preservation, some are missing and others are broken. There is some evidence of xylophagous insects (termites) with action in one of the scissors, especially in the one located next to the wooden partition.

Besides the walls that go around the construction, at the limits of Façades E1 and E3, wooden supports that are launched are thrown, in the extension of the scissors, still original, that protect and compose the two external circulations with cement

Fig. 9 Detail of the openings and barred, next to Façade E1, (facing east), in the foreground we can partially see the masonry cloth, previously totally blind, where a vitro and door was implanted. Photo of the author

floor, one of them over where the main access was given, next to the Façade E1, and the other, over the platform, where the passengers board and disembark. This last one was elevated in relation to the still existing rails. The support of French hands is given next to the walls of the façades mentioned above, however, on masonry pilasters that modulate the whole space of the station.

The French hands of Façade E3 were roughly sawn, in all alignment of its eaves, to give shelter to compositions that transport grains, very high, that began to park there in the late 1970s. This information came to us from Mr. Santos, a retired carpenter from the old RFFSA, even more so in the active one, who was precisely one of those who sawed the set in 1978. It is true, also supported by old photographs, that originally the roofs next to the French hands of the two façades had the same dimension and design, and the main support joists of the same counted, in its tailpiece, with pigeon breast finish, disappeared in Façade E3, however, it is still found in Façade E1.

It is verified that the structures of support of the French hands more compromised by the humidity are those in the two extremities of the construction, due to the breaking of the tiles and also of the easy reach of the rains, such pieces precision to be remade (Fig. 10).

Fig. 10 Internal view of the roof and wooden frames. It is perceived the bad state of the original tiles. Photo of the author

3   Preliminary Guidelines for a Restoration Project

One of the clearest findings, as exposed in the diagnosis about the station building, is that it has lost its original characteristics and that trying to recover this initial identity would be something unreasonable, false, and unnecessary. The inexistence of the original project, internal photos, or even reliable reports about the station, disallows its internal reconstitution with the original language, which could become an unreasonable pastiche. When the elevations, because a larger amount of images were found and the marks of the changes were "registered" in the apparent masonry walls.

The fact that the station no longer has the internal walls and received another configuration of space in the 1940s points to a more prudent and less fanciful future restoration, on the return of its first configuration, however, at the same time, due to this same situation, allows a greater freedom of project/creative action. The Charter of Venice is clear about the conditions that must order a good restoration work:

"It ends where the hypothesis begins; in the plan of conjecture reconstitutions, all complementary work recognized as indispensable for aesthetic or technical reasons will stand out from the architectural composition and should bear the mark of our time. The restoration will always be preceded and accompanied by an archaeological and historical study of the monument". (ICOMOS, Letter of Venice 1964; IPHAN: Notebook of Documents No. 3: Heritage Letters 1995, pp. 110–111).

In an eventual restoration proposal, a safe path must be followed in relation to the construction, intervening only where there can be certainty and good grounds regarding the characteristics of the good.

In addition to the aspects listed above, an eventual restoration proposal must contain:

- Adaptations that can be limited to a minimum and with the possibility of reversibility;
- Restoration of constructive and architectural elements using contemporary techniques recognized and recommended by specialists;
- Restoration in order to guarantee the durability and the integral use of the good, avoiding or delaying its degradation;
- Restoration that foresees and makes clear the distinctiveness of the old materials from the current ones. Where there is a situation of replacement of original elements (woodwork, frames, etc.), by others of the same design, these should be clearly distinct through painting in a different tone from the authentic ones. We also indicate that the use of the replica should be expressed in a plate to be fixed in the building.
- Restoration that foresees and incorporates fair and contemporary demands such as universal accessibility, patrimonial security, new media, flexibility, and mutability in the use of spaces.

4   Final Considerations

The preservation of the old Sorocabana Company station is of utmost importance, due to its role in the growth and enrichment of the city. It was the gateway for the coming of two other railroads that transformed the city into a rich regional hub for commerce and services, which currently has a population of approximately 400 thousand people and has the 68th gross domestic product (GDP), of Brazil.

The overturning of Bauru's railroad finally approved in 2017 by the state preservation agency, CONDEPHAAT, facilitates the preservation process due to the terms of the law that prevent its demolition and new features.

Finally, this work tried to show the historical relief of the station to the city of Bauru and briefly point out the diagnosis of the physical situation of the building, held in 2011. This was one of the initial steps for the executive restoration project, completed a year later, but not yet implemented. It remains now to get resources so that the space can finally be restored, with the purpose of receiving public use.

References

Brandi CT (2004) Translation by Beatriz Mugayar Kulh. Ateliê, Cotia

Caderno de resumos do VI Colóquio Latinoamericano sobre recuperação e preservação do Patrimônio Industrial. IV Encontro Internacional sobre Patrimônio Ferroviário. São Paulo, Centro Universitário Belas Artes

Conselho de Defesa do Patrimonio Histórico, Arqueológico, Artístico e Turístico do Estado. Processo n° 30367 Assunto: Solicita o Tombamento de edifícios pertencentes a sede da Antiga Estrada de Ferro Noroeste do Brasil, situados em Bauru

Conselho Internacional de monumentos e sitios ICOMOS. Carta de Veneza, 1964. In: Iphan: Cartas Patrimoniais. 3ª Ed. Ver. Iphan, Rio de Janeiro, pp 91–95

Decree n 373 - 15 de Julho de 1896, Companhia Paulista de Vias Férreas e Fluviais, Arquivo Câmara Municipal de Bauru

Ghirardello N (1992) Aspectos do Direcionamento Urbano da Cidade de Bauru. São Carlos, 1992, 187 p.. Unpublished Master Thesis (Arquitetura e Urbanismo) - Escola de Engenharia de São Carlos, Universidade de São Paulo

Ghirardello N (2002) À Beira da Linha, Formações urbanas da Noroeste paulista. Editora da Unesp, São Paulo

Ghirardello N (2010) 2010. A formação dos patrimônios religiosos no processo de expansão urbana paulista, São Paulo, Editora da Unesp

Governo da Italia. Carta de Restauro, 1972. In: IPHAN: Cartas Patrimoniais. 3 Ed.Ver. Rio de Janeiro: Iphan, 2004, pp 91–95. In: IPHAN: Cartas Patrimoniais. 3ª Ed. Ver. Rio de Janeiro: Iphan, 2004, pp 147–169

IPHAN - Instituto do Patrimônio Histórico e Artístico Nacional (Brasil). Cartas Patrimoniais - Brasília: IPHAN, (Cadernos de Documentos n° 3)

Kulh BM (1998) Arquitetura do ferro e arquitetura ferroviária em São Paulo: reflexões sobre a sua preservação. São Paulo: Ateliê: Fapesp: Secretaria da Cultura

Manual de Elaboraçao de Projetos de preservação do patrimônio cultural.GT/IPHAN. Brasília, junho de

Matos ON (1974) Café e Ferrovias: A Evolução Ferroviária de São Paulo e o Desenvolvimento da Cultura Cafeeira. Alfa-Omega. Sociologia e Política, São Paulo

Normas de Projetos (1980) IEPHA/MG, Superintendência de Conservação e Restauração – Setor de Projetos. Belo Horizonte

Pauleto LSTL (2006) Diretrizes para intervenções em edificações ferroviárias de interesse histórico no Estado de São Paulo: as estações da Estrada de Ferro Noroeste do Brasil. Unpoublished Master Thesis. Faculdade de Arquitetura e Urbanismo, Universidade de São Paulo

Pérez F (1918) Álbum illustrado da Companhia Paulista de Estradas de Ferro – 1868-1918. s. n, São Paulo

Sites

http://www.estacoesferroviarias.com.br/b/bauru-efs.htm
http://www.sp-turismo.com/mapas

Heritage and Museums: The Cultural Significance of the University

Paulo Henrique Martinez

Abstract A reflection on the impacts of the social and cultural changes of the second half of the twentieth century on the daily work with heritage and museums at the Brazilian university. The proposition of the United Nations social agenda from the 1990s to 2030 created possibilities for the development of individual, institutional, and social capacities in teaching, research, university extension, and cultural actions from heritage and university museums.

Keywords Museum · Heritage · University · Brazil

Let us take a fact known to most Brazilians, the television series "The Big Family". It has been on the show for over a decade, between 2001 and 2014, being the best known and most accessible version for several generations. Those in their 50s should remember the first edition of the program, in the 1970s, shown between 1972 and 1975. It is from this first phase that I will begin a professional reflection.

The program had become a success of public and criticism, a metaphor of the Brazilian nation itself. There was no one who did not identify in the characters, scenarios, behaviors, lines, clothes, accessories, decoration, situations and expectations, any close relative someone known, when not himself/herself. It was the greatest success in the national comedy of Brazilian television customs and a moment of a singular encounter of art with society, in the middle of the military dictatorship. It would soon become evident that the author of the successful formula was Oduvaldo Vianna Filho, Vianinha. It was he who wrote the texts of the episodes from April 1973 onward, and who was still responsible for setting and defining the daily themes of a suburban family, but who reached the immense majority of Brazilian families, from the middle class downward, by portraying the rivalries, solidarity, and perseverance in facing the hardships of life. Vianinha died in July 1974, a victim of cancer, at the age of 38. He was succeeded in the creation of the program by Paulo Pontes, another talent of the same generation of Brazilian dramaturgy, but the texts did not have the

P. H. Martinez (✉)
State University of São Paulo "Julio de Mesquita Filho" (UNESP), Sao Paulo, Brazil
e-mail: ph.martinez@unesp.br

© The Author(s), under exclusive license to Springer Nature Switzerland AG 2021   187
F. Lopes da Cunha and J. Rabassa (eds.), *Festivals and Heritage in Latin America*,
The Latin American Studies Book Series,
https://doi.org/10.1007/978-3-030-67985-9_11

same success, and "The Big Family" went off the air in March 1975. Paulo Pontes himself died in 1976, also a victim of cancer, at only 36.

What matters to us here, at this moment, is to recover the observation made in an interview that Vianinha granted some months before his death. The interview was never formally published and some passages were made known in the obituary published by Visão magazine, in August 1974. The excerpt I want to highlight alluded fundamentally to the following: "The Brazilians need to look in the eye the tragedy of his country". And further: "to look into the eyes of tragedy is to make it dominated". Vianinha made reference to the degree of spoliation, exclusion, and cultural alienation, characteristic and predominant in Brazilian society. To achieve this tragic effect, he said, oppression and income inequality, violence, poverty, dictatorship are not enough. It takes "a very intense and sophisticated cultural process" to "brutalize this society," which only the technology of mass media, advertising, and urban landscaping could ensure to hide a tragic reality that reduces a society of millions of people to a consumer market that represents a quarter of that population. The domestication of three quarters of the population requires very refined and elaborate strategies (Viana Filho 1999).

Vianinha's death made it impossible for us to have the continuity and deepening of his reflection on cultural processes in Brazil. I return to this thinking to meditate about the cultural significance of the public university. I understand that this cultural meaning of the university lies in triggering processes of cultural deepening to de-embruit the Brazilian society. And here, technical and scientific rationality, utilitarianism, and productivism will not be enough and do not even have the purpose of operating this change. Cultural processes should be deeper, because the wound bleeds internally, at the core of subjectivity, emotions, sensations, and affections. It will not be possible to touch it, treat it, suture it, heal it, without high doses of solidarity, arts, and humanities (Boaventura 2010).

It is necessary to trigger a broader and more powerful cultural process. This is what Vianinha's reflection suggests to us and what I hope to demonstrate next. All the more so than rendering accounts, the public university must propose ways and answers to face the contemporary challenges of Brazilian society. Explaining the cultural significance of the public university in Brazil involves both a critical understanding of its past trajectory and the formulation of a social project for its insertion into twenty-first century society. Every social project starts, necessarily and inevitably, from a social memory (Fontana 2013).

The perspectives envisioned in Agenda 2030 for the world society and that the proposition of the Sustainable Development Objectives intends to rectify, until then, are of a more urbanized life, highly technified, greater population density, culturally diversified, regional and socially unequal, violent, strangled by the environmental crisis and its global and concrete effects (climate change, pollution, water shortage, loss of biodiversity, food insecurity, epidemics, migration, among others). A maximum of contrasting humanity with a minimum of humanism and solidarity. All the institutional action of the university should reach and transform, in some measure, the reality of daily school life in public education. It is there that the rawest interface, in living flesh, of inequality and social violence in Brazil lies. The beginning, the end and the middle of the perverse spoliation, exclusion and cultural alienation in our society.

It is not fit here to make an inventory of the tragedy of the programmed and consented ruin of the public university lived and known for a long time. Every crisis is the starting point of a new beginning. The public university in Brazil, born in the 1930s, was installed with the purpose of operating as an instrument of political affirmation and social domestication through education for industrial work and technological modernization. In the social and economic conditions prevailing in the country, the Brazilian university quickly became an instrument of social and national emancipation, extrapolating the initial intention of its creators. The case of the imprisonment of students from the social sciences course at the University of São Paulo, as early as 1934, when they were conducting field research in the neighborhoods of the capital city of São Paulo, when the sociologist and socialist names seemed synonymous to the police authorities, and the closure of the University of the Federal District in 1935, are emblematic episodes. The coup d'état in April 1964 reopened the cycle of open violence against Brazilian universities.

In Brazil, the public university has become a political problem since its birth. A political problem that persists today and that will only be solved by the full and effective institutional realization of the public, creative university, as it can and should be, or by its submission to the onslaughts of siege and annihilation over time. The latter were systematically accentuated in the contests against the Public School Defense Campaign in the early 1960s, in the dismantling of the institutional and pedagogical project of the University of Brasilia (UnB) after 1964, and in the university reform of 1968, with the castration of teachers and public servants promoted by the military dictatorship in 1969. The so-called crisis of the university emerges as a lasting social process and the onslaughts are reaffirmed with the intensification of neoliberal policies during the 1990s, with the end of the Cold War and the so-called crisis of utopias. The consequences are quite concrete in social and university life today.

A new intellectual, institutional, and political engagement can be built in response to the new cultural needs of the twenty-first century. An "open and democratic university, fused with the interests of the majority, that is, the mass of the poor and working population" (Fernandes 1984, 20). Finally, I want to point out three possible strategies for university innovation and the transition from conformism and non-conformism to active interaction, giving social and cultural meaning to the university experience for established and outsiders, those who will tread their lives inside and outside the university institution.

The first strategy must be focused on the search for social and cultural innovation through the promotion of digital culture, arts, and humanities. It is focused on the social valorization of young people, in an effective, open, and intense alliance with the schools of the official educational network, municipal and state. The second strategy should stimulate institutional innovation in the promotion of inter-institutional and international academic cooperation, founded on research and the solution of twenty-first century world problems. It focuses on the social valorization of cultural, ethnic, and gender diversity, particularly in the daily work at the university. The third strategy contemplates political innovation for the promotion of human rights, citizenship, democracy, and the formulation, implementation, and evaluation of public policies. It focuses on the social valorization of regional communities with greater proximity, participation, and integration to university campuses, inducers of the economy of culture, based on creativity. Some ways to trigger these strategies can be glimpsed right now.

1 The Age of Institutions

The end of World War II opened the way for the growing division of intellectual and scientific work, accompanying the changes in the sphere of economic production, particularly in industry and continuous technological improvement. The new international economic moment corresponded to a new moment in the organization and functioning of institutions, public and private, in all spheres of social life. The most emblematic and recognized of these institutions that emerged from 1945 onward is undeniably the United Nations (UN), created that same year. UNESCO was then created to promote coordinated actions and international dialogues in rebuilding and building educational, scientific, and cultural structures within the new community of nations.

In the field of culture, the institutional dynamics operated in front of two general trends, gradual and progressively consolidated. One of them was the rapid growth of the cultural industry, notably expressed in the massive advent of advertising and television. Another trend was the massification of access to basic and university education. These transformations took impulse in Western European countries, such as France, England, Italy, and Germany, in Scandinavian countries, North America, Canada, and the United States. On a smaller scale, they reached Latin American countries such as Argentina, Brazil, Mexico, and Venezuela, and Asian countries such as India and Japan.

Structural changes in the industrial and technological economy have brought new dynamics to social relations, imposing patterns of consumption of goods and services as a way of life. There has been a gradual change from an ethic of manual and industrial work to an ethic of consumption which, in many circumstances, has contaminated family relations and between different generations, for example, under the misunderstanding of meanings, value, attachment, and detachment, to professional activities and the very conception of what work was. In the field of arts and culture, the American critic, Marshall Berman, highlighted the separation between aesthetic and political ideals of the modernist vanguards and the rapid and intense changes brought about under the motto of modernization of a technical and industrial nature. The privileged terrain of the occurrence and public visibility, advertising, of these transformations was the urban space of the big cities and its repercussion in small cities, in the ardent desire of the smaller ones to be, in the future, like the bigger ones (Berman 1986).

It was in the United States that another American scholar, Russell Jacoby, identified, in the industrial city after 1945, the recurrent signs of social change that would have disrupted known urban environments in daily life, such as cafes, bookstores, small theaters and arts, and entertainment halls. Likewise, there were changes in habits of bourgeois sociability, such as flanerie—the walks to enjoy and appreciate urban landscapes—and bohemian coexistence, as well as in languages of literate culture, books, newspapers, and magazines. The flaming cultural life that pulsed in the streets and neighborhoods of large cities would have been shaken by the rise of society and the cities of the automobile. Express roads, bridges and viaducts, speed and individual transportation, and shopping centers changed the circulation of pedestrians and the gathering of small groups, imposing new rhythms to cultural and social life, with the character of mass production and consumption (Giucci 2004; Caruso 2010). The other side of this coin of social and urban transformation was stamped on university cities. These contained other organized spaces, planned or not, enclosures, and environments specialized in the production of social and cultural meanings. The university cities and college campuses that would be installed in the outskirts and suburbs, with distant access by roads and highways, emptied bohemian life and cultural routines, spaces for artists and intellectuals to live together and perform in the cities.

Within universities, the role derived from this new cultural condition in social production and technification, as an infinite necessity, has given way to widespread disorientation, particularly in the arts, social sciences, and humanities (Lledó 2018). The rapid and intense expansion of professionalization and specialization in intellectual work has been noted both in basic and university education and in other public and private scientific and cultural institutions. In the 1960s, cultural tension overflowed from university campuses and other educational spaces to the streets, student demonstrations and social mobilization boomed in different countries. Today, when observed in historical perspectives, of a few decades and generations, the salaried and competitions in the university system, the specializations, the formative dynamics of students—serial disciplines, internships, scientific initiation, graduate programs, post-doctoral, scholarships, monographs, disciplines, research groups, publications—reveal the insularity of the work of individuals, institutions, and entire areas of knowledge.

The loss of vitality of public culture translates the loss of vitality of public education. The fragmentation and dispersion of cultural life that have gained expression in the deconcentration of cities and the displacement to university campuses have contributed little to public culture. This finding allowed the observation of what Russell Jacoby called an eclipse of public intellectuals. The intellectuals. The fact that the nonacademic world would have become extinct due to the devastation of its urban cultural environments and the undisguised existence of academic intellectuals who are not fully integrated into the university systems in force today inside and outside Brazil (Jacoby 1990, 19).

The changes in public culture do not mean that it has disappeared. The school environment, a privileged space for literate culture, and the official educational network, with its thousands of students, teachers, and professionals, are home to a large number of unassisted citizens in Brazilian society. Basic education, public schools, students from neighborhoods, central and peripheral areas, and poor communities should be the priority in the cultural action of the public university. Only the university can assure this segment of the Brazilian population the initial and continuing education for insertion, identification, and exploitation of talents and professional opportunities in this century. Only the public university will be able to provide for the difficulties of attention and concentration of a non-readership in activities of literacy and scientific dissemination, of university and cultural extension. Public assets also make possible countless approaches aimed at the lay public, extending the radius of action of universities to non-formal education.

2 The New International Agenda and Capacity Building

The closing of the ideological bipolarity of the Cold War opened political and diplomatic space for the search for a new world consensus. The 1990s were marked by successive meetings within the United Nations, which had as their starting point the promotion of human rights, democracy, and human development. The centrality of international action and national public policies should reside in people and attention to the most vulnerable social groups in each society, indigenous peoples and migrants, and women, the elderly, children. The World Conferences were dedicated to diagnoses and the establishment of action plans on social themes such as environment (Rio de Janeiro 1992), human rights (Vienna 1993), population (Cairo 1994), social development (Copenhagen 1995), condition of women (Beijing 1995), and urban life (Istanbul 1996).

International cooperation should replace the logic of ideological confrontation and the social agenda proposed by the UN sought to establish parameters for coexistence in the world order of the nascent twenty-first century. The complex social problems, shared globally, should be faced jointly by the international community. The humanist approach appreciated by the UN highlighted the growing interdependence of social and individual life, placing at the center of the debates the necessary conditions to assure the population, on a global scale, quality of life, access to health, education, housing, employment, and income. World economic growth should be associated with the search for greater equity among countries and within national societies (Sachs 2007).

The valorization of human beings as subjects of rights, the stimulation of self-development, and the observance of human rights sought to give public visibility and make effective the approach to social issues in their interrelation with the global agenda for sustainable development. This perspective sought to integrate social issues into the environment, human rights, the situation and role of women, in different societies and cultures, urban life, and social development. The global and integrated perception made evident a general objective to be achieved, on a global scale, in the twenty-first century: the eradication of poverty. The World Conference for Social Development, held in Copenhagen, was an unprecedented meeting in the history of international relations. The Conference included social issues in the list of global issues that cannot be postponed for political and economic stability in the twenty-first century. Observed together, the social agenda of UN international meetings in the 1990s resulted in diagnoses and the definition of joint action guidelines and goals (Rubarth 1999).

In 2000, the proposition of the Millennium Development Goals aimed at the global confrontation of world problems, with the eradication of poverty as one of the eight goals related to the eradication of hunger and poverty—valorization of women, diseases, education, environment, pregnancy and infant mortality, cooperation, and partnerships—to be achieved by 2015. In sequence, this same year, the UN social agenda was renewed and expanded with the proposition of the Sustainable Development Goals, establishing goals and guidelines for their achievement by 2030.

The holding of periodic general and thematic meetings and the mobilization of governments and society have characterized the actions of UN bodies since its creation in 1945. The elaboration of the social agenda promoted in the 1990s recommended new approaches between the public sector and society, the market and non-state subjects, in the formulation of general and sectorial policies and the constitution of inter-institutional partnerships and varied scales, from local to global. It was in this context that universities, the scientific community, the media, and non-governmental organizations of all shades were called to engage in achieving the objectives and goals defined in declarations and action plans of the United Nations social agenda. An arena of debate and reflection was opened for a cultural agenda under globalization and integrated with the purposes of sustainable development. UNESCO launched the Decade of Education for Sustainable Development between 2005 and 2014, mobilizing university institutions for a worldwide commitment in this direction (UNESCO 2015).

All this effort at the international level has been harshly confronted with the rise of neoliberally inspired economic policies, as well as new and resurrected old rivalries between nations, the Persian Gulf wars, and civil wars such as those in Europe, Africa, and Latin America. Structural reforms of the global and international economy have adopted plans to retract state action, especially in social issues, eroding many of the foundations for the UN social agenda. The reduction of public investment, policies to protect labor conditions and relations, job and income generation, restrictions on social support and state subsidies in different branches of economic activity, fiscal adjustments, among other measures, have aggravated the social effects on the world and national economies (Touraine 2011). Successive financial and global crises have ruined the willingness and readiness for public policies aimed at the goals and targets of the Millennium and Sustainable Development, proposed by world conferences over the past two decades. The effects of the 2008–2009 crisis are present today, among them the persistent economic recession and the two pandemics experienced in this century, that of the influenza A H1N1 in 2009 and the current Covid-19.

There is an ongoing restructuring in the field of cultural activity that was undressed under that global crisis. Culture has been thought, from the perspective of integration to sustainable social development, as a field of opportunities for the generation of

direct jobs and investments in the publishing and audiovisual industry, digital technologies, tourism, tax collection, and international currency. Cultural activities foster
a creative and cultural economy, linked to the management of goods and services, and
a natural and cultural heritage economy, under the inspiration of the heritage policy
recommended by UNESCO. It is in the search for making these proposals of the UN
social agenda effective, and reaching those in the field of culture through UNESCO's
formulations and guidelines, that the processes of acquisition, sharing, and diffusion
of information and knowledge become relevant (de Cuéllar PJ (Org.) 1997). In the
assessment of Carlos Lopes, new forms of social exclusion may derive from existing
fragilities in countries, which inhibit their entry and the use of eventual opportunities
of insertion into the world economy and the exercise of global citizenship (Lopes
2005).

Carlos Lopes proposed the concept of capacity development as a methodological
guideline for work in multiple dimensions and in the articulation, in a systemic vision,
of integrating strategies of social issues to the objectives of sustainable development,
in its variants of time and space. The development of capacities is here incorporated
in the expectation of improvement of social, individual and collective, and institutional performance. The concept involves both the goal dimension to be reached
and the constructive practice dimension in the planning, monitoring, and evaluation
of different processes, initiatives, and actions from technicians to specialists, from
governments to universities, from NGOs to companies, and from schools to cultural
institutions.

The emphasis of capacity development is on strengthening existing capacities.
These capacities are activated from three articulated dimensions. The first, individual,
promotes the improvement of people's skills and abilities; the second, institutions,
seeks the improvement of structures and operational actions; and the third, social,
involves the understanding of diversity and the expansion of opportunities that open
up in the performance of public and private sectors and the so-called third sector.
The development of capacities aims at the definition and perception of objectives to
be achieved. The identification of problems and the confrontation of challenges in an
effective and sustainable way involve the appropriation of concepts, processes and
methods, cooperation, partnership, organization, interpersonal, inter-institutional and
international relations, responsibility, behavior, values, legislation, public policies,
and political and social participation. Capacity building associates the participation of
people, institutions, communities, and countries to the knowledge and interpretation
of the world, without ignoring that access to education, science, culture, and technology, defines the levels of quality, intensity, frequency, and reach in the promotion
of human rights, citizenship, and democracy.

3 University, Heritage, and Museums

In our universities, there is a diverse set of heritage, natural and cultural, museums, and science centers. Despite their intrinsic value in the national memory and in the institutional and cultural history of Brazilian society, these heritages also represent opportunities in the generation of new knowledge and in the public access to the inventive imagination, in its scientific and technological dimensions, and in the valorization of the human being by the development of capacities, in its political and cultural dimensions, in the performance of universities in the twenty-first century. There is a universe of pedagogical, methodological, empirical, theoretical, and applied research practices, which provides opportunities that are little or not taken advantage of, in countless situations, in the construction of public knowledge, in initial and continued training and professionalization, in different areas, disciplines, and interdisciplinary approaches. The work with diversified typology of heritage, collections, natural and cultural assets is multidimensional and fruitful in the exercise and experimentation of the indissociability between teaching, research, and university extension.

This fact, the existence of significant heritage and museum collections in Brazilian universities, especially public ones, requires institutional policies that effectively ensure the right to culture and social uses of university heritage. These are both public property and global public goods that, under custody, are destined for preservation and dissemination by universities for the benefit of society as a whole. A public service to be rendered, that of promoting social memory, natural and cultural heritage, tangible and intangible (Kaul 2012).

In the twenty-first century, we have witnessed some tragedies in the field of heritage and museums in Brazilian universities. The headquarters building of the Paulista Museum of the University of São Paulo (USP), known as the Ipiranga Museum, has been closed for almost a decade. The National Museum of the Federal University of Rio de Janeiro (UFRJ) was consumed by the fire in 2018, the year of the bicentennial of its creation. The Museum of Natural History of the Federal University of Minas Gerais (UFMG), in Belo Horizonte, on a smaller scale in losses and destruction, had a similar fate in 2020. The frequent losses, conscious or not, thefts, deterioration, negligence, neglect, destruction, and forgetfulness of heritage and museums in universities are revealing. They reveal the insular existence in which the public university finds itself and the historical need for constant remodeling of intellectual work and cultural action in the public university. They also point out that by not meeting the primary imperatives of promoting heritage and university museums, our universities fail to provide a public service and ensure a constitutional right, the right to culture. Article 215 of the Brazilian Constitution of 1988 says: "The State will guarantee to all the full exercise of cultural rights and access to the sources of national culture, and will support and encourage the valorization and diffusion of cultural manifestations" (Brazil 1988, 141). Thirdly, the observance of the prerogatives of the right to culture is a very appropriate conduct to affirm the promotion of creative work in the field of heritage and museum collections existing today in our university institutions.

In short, little or insufficient attention to heritage and university museums has a perverse effect: (1) the accelerated and prolonged loss and deterioration of these assets; (2) by making them inaccessible to knowledge and public enjoyment, the assets are ignored and, therefore, no longer valued by society and lose interest in their existence, apparently devoid of meaning; (3) the university itself often takes care of heritage and museums as something without relevant public function, delivering heritage and museums to its own fate, in the absence of specific public policies, endowed with regular institutionality, continuity, and financing. There are exceptions, it is true, but the selfless, generous, and voluntary action of public servants, teachers, and students in universities ends up prevailing. In spite of countless, laudable, and tireless initiatives, the discontinuity in management, the precariousness of actions, and the scientific, pedagogical, and cultural immobilism are reproduced indefinitely.

The weakening of the public university before society is quickly installed in collective incomprehension, political attacks and financial and budgetary cuts, made by agents of the market of education and culture, of conservative, secular and religious thought, in the voice of their corporate, partisan, and ideological representatives, contributing to a distorted and incomplete perception of the cultural meanings of the public university in general and of the university patrimony and museums in particular. The effects are well known. In Brazil, starting in 2016, we are living under the deliberate and undisguised institutional emptying of those meanings in the collective orientation and in the actions aimed at dismantling social relations in national life and their impact in the daily political, educational, scientific, and cultural life of the immense majority of the population.

A culture of heritage and museums, like any other artistic language and scientific thought, is also a public culture, necessary and inescapable, political. The consolidation of a culture of heritage and museums, as a public culture, requires both critical thinking and scientific and technological specialization. There is no public policy without adequate provision of theoretical, conceptual, technical, and pedagogical goals. The effective right to culture, the construction of new knowledge, and the formation of creative and qualified professionals in Brazilian public universities cannot abstract this institutional requirement. The emptying of culture into public life does not eliminate its critical and transformative role in Brazilian public universities.

Campuses and university cities and their built and natural, material and immaterial heritage are stimulating territorial units and starting points for diverse initiatives to promote capacity development. Seen from an integrated perspective, territory and heritage, they allow one to get to know the structures and daily life of coexistence relationships and teaching and learning situations in which the social role of the public university becomes more evident and present in the surrounding society. The heritage of the university territory—campus, campuses, university cities—becomes a learning community when, by activating its cultural and natural assets, existing or related to them, it triggers a process of appreciation of local history and comparison with other scales of time and space of the University and social and cultural life (Starling and Duarte 2009; Borde and Bellinha 2015). The promotion of dialogues on practices and senses of citizenship, human rights, and democracy strengthen the social, educational, and cultural development of different audiences and communities, neighborhoods, and schools.

The growing acceptance and dissemination of these guidelines are the result, on the one hand, of the innumerable and followed findings, in different situations and possibilities opened in the access to knowledge, its production and construction, by the dialogue between the university and non-university knowledge. The debate on the role of museums in promoting development and social change, on the other hand, gained momentum in 1972 with the questions that laid the foundations for the proposal of a new museology. Twenty years later, a set of practices and concepts that guide the so-called Social Museology (Bruno MCO (Coord.) 2010) took shape.

Some years ago, the Federal University of Rio Grande do Sul (UFRGS), for example, developed projects that started from formulations of Social Museology, in search of the effectiveness of the strategic indissociability between teaching, research, and university extension. The generation of knowledge of a new kind, built in the development of activities of approximation between reason and emotions, with the participation of residents of peripheral communities around Porto Alegre, provides an example of behavior and attitude of the public university toward the heritage, territories, and populations exposed to conditions of greater social, educational, and cultural vulnerability. Memory wheels, a street museum, a heritage education workshop, and a community tourism route were some of the tools developed jointly with the residents and schools (Zen (Org.) 2016).

A circuit of dialogues was operated by the articulation of personal narratives, daily relationships, and the meanings and social reach of forms of life, human and non-human, in dozens of neighborhoods that integrate the communities involved in the projects of that University. The territories, the natural and cultural heritage, tangible and intangible, and the local history of settlement, occupation, and use of land, urbanization, quality of life and expectations for the future insert spaces and different social groups in the time of the world and national life. Individuals and collectives, aware of their history, cultural identities, and political options, emerge.

Education for heritage, the proposal of cultural mediation that guided the triggering of the different activities, has enabled UFRGS university students and residents participating in the projects to assume the exercise of rights and responsibilities, in different initiatives of interest to the entire population and their respective communities (Varine 2013). The concept of ownership has become effective, one of the instruments in capacity development. By becoming active and conscious subjects in the exercise of individual, institutional, and social rights and responsibilities, the community has also known the acquisition, control, and understanding of the condition of agents of its wills and interests in decision making and action choices in the achievement of desired and expected goals and objectives. The concept of empowerment has become effective, this other instrument in capacity development (Lopes 2005).

4 Conclusions

The resignification and the possibilities open to different cultural activities in recent decades have not only involved universities but have also found there privileged spaces for their elaboration and social dissemination. New cultural, professional, and technically sophisticated dynamics have emerged in forms of production, organization, management, and consumption of cultural goods and services. These dynamics have reached both public and private institutions of education and culture as well as the cultural industry and mass consumption (Canclini 2006).

The proposal of capacity building oriented to social and cultural innovation, institutional and political is a stimulating and challenging strategy for imagination and critical thinking in natural and cultural heritage spaces, museums, and science centers. These are existing or potential spaces in campuses and university cities that harbor opportunities for the exercise of indissociability between teaching, research, and university extension and cultural actions and that can give impetus to that initial proposition. Nowadays, education and public knowledge, technical and professional improvement, other forms of organization, participation and political decision making, can be articulated in the experimentation and implementation of public policies in solidarity and innovative cultural meanings in our universities.

In the era of institutions, the obsolescence of the economic, industrial, and technological park has aroused more interest than the intellectual park. After World War II, U.S. governments and universities, for example, focused their attention, investments, and infrastructure on training and hiring specialists and technicians, not on the creative and inventive imagination of artists and intellectuals, activists, and independent thinkers (Jacoby 1990). In promoting the social agenda for the twenty-first century, the ideas of solidarity and induced and sustainable changes find in the development of capacities, from the universities, strategies to initiate the unavoidable process of unpacking social relations and cultural meanings of the public university in Brazil.

References

Berman M (1986) Tudo o que é sólido desmancha no ar: a aventura da modernidade (transl by Moisés CF, Ioriatti AML). Companhia das Letras, São Paulo

Borde ALP, Bellinha PRT (2015) (Org.) Conservação e reativação do patrimônio arquitetônico universitário. Prourb, Rio de Janeiro

Brasil (1988) Constituição: República Federativa do Brasil. Senado Federal, Brasília

Bruno MCO (Coord.) (2010) O ICOM-Brasil e o pensamento museológico brasileiro: documentos selecionados. São Paulo: Pinacoteca do Estado; Secretaria de Estado da Cultura; Comitê Brasileiro do Conselho Internacional de Museus (2 vols.)

Canclini NG (2006) Consumidores e cidadãos (transl by Dias MS), 6º edn. UFRJ, Rio de Janeiro

Carnoy M (2004) A educação na América atina está preparando sua força de trabalho para as economias do século XXI?. UNESCO, Brasília

Caruso RC (2010) O automóvel: o planejamento urbano e a crise das cidades. Officio, Florianópolis de Cuéllar PJ (Org.) (1997) Nossa diversidade criadora, transl by Candeas AW. UNESCO, Campinas: Papirus, Brasília

Fernandes F (1984) A questão da USP. Brasiliense, São Paulo

Fontana J (2013) El futuro es un país extraño. Passado & Presente, Barcelona

Giucci G (2004) A vida cultural do automóvel: percursos da modernidade cinética, transl by Martins A. Civilização Brasileira, Rio de Janeiro

Jacoby R (1990) Os últimos intelectuais: a cultura americana na era da academia (transl by Lopes M). Trajetória; Edusp, São Paulo

Kaul I et al (2012) Bens públicos globais: cooperação internacional no século XXI (transl by Maldonado Z). Record, Rio de Janeiro

Lledó E (2018) Sobre la educación. Taurus, Barcelona

Lopes C (2005) Cooperação e desenvolvimento humano: a agenda emergente para o próximo milênio. Unesp, São Paulo

Rubarth EO (1999) A diplomacia brasileira e os temas sociais: o caso da saúde. Instituto Rio Branco, Brasília

Sachs I (2007) A terceira margem, transl by D'Aguiar RF. Companhia das Letras, São Paulo

Santos BS (2010) 3° edition. A universidade no século XXI. Cortez, São Paulo

Starling HMM, Regina Duarte (Org.) H (2009) Cidade Universitária da UFMG: história e natureza. Belo Horizonte, UFMG

Touraine A (2011) Após a crise (transl by Morás F. Vozes, Petrópolis)

UNESCO (2015) De ideias a ações: 70 anos da UNESCO (transl by Teixeira G.) Santos: Editora Brasileira de Arte e Cultura. UNESCO, Paris

Varine H (2013) As raízes do futuro: o patrimônio a serviço de desenvolvimento local, transl by Horta MLP. Medianiz, Porto Alegre

Viana Filho O (1999) Vianinha: teatro, televisão, política. Brasiliense (Org. F. Peixoto), São Paulo

Zen Dalla (Org.) AM (2016) Aulas de Museu. UFRGS, Porto Alegre

Tropas and *Tropeiros* in Southern Brazil: History, Memory and Heritage

Milena Santos Mayer and Fabiana Lopes da Cunha

Abstract This article is the result of a dialogue between the authors based on a doctoral research in History, still in progress, which has as its central object the trajectory of a museological institution called *Museu do Tropeiro* (Museum of Tropeiro). This museum is located in the interior of Brazil, in the city of Castro (Paraná) and is dedicated to the preservation and dissemination of the history and memory of the mules' trade in the southern region and its social and cultural implications in the municipality and the region. This was an activity that began in the colonial period; when the need arose to transport cargo and beef animals throughout the Brazilian territory. This practice can be evaluated as a global phenomenon since the use of animals was for a long time the main means of transportation for humanity. However, in southern Brazil, this activity has developed with the peculiarity of the significant commercialization of mules. These animals were transported from the region of the pampas to the city of Sorocaba, located today in the state of São Paulo, and then sold to be used elsewhere in the country as means of transportation for people and goods. In this way, long roads were built that made possible the integration of a part of the Brazilian territory that was far from the relatively known coast. The *tropeiros* (or muleteers), men who drove and traded these animals, had the need to stay overnight in certain places for their own rest and for the reestablishment of the *tropas* (or trains). One of the main stopping points was the Campos Gerais region in the current state of Paraná, propitious for its field vegetation, the region has as its oldest administrative organization the current municipality of Castro. In 1977, when the city conquered its public museum, it emerged as a thematic museum, the first in the country dedicated

M. S. Mayer (✉) · F. L. da Cunha
Paulista State University Júlio de Mesquita Filho (UNESP), Researcher financed Coordination for the Improvement of Higher Level Personnel (CAPES), Assis, São Paulo, Brazil
e-mail: milenasmayer@gmail.com

Paulista State University, Julio de Mesquita Filho, São Paulo, Brazil

F. L. da Cunha
e-mail: fabiana.cunha@unesp.br

M. S. Mayer
Researcher financed by CAPES Foundation, CAPES, Sao Paulo, Brazil

© The Author(s), under exclusive license to Springer Nature Switzerland AG 2021
F. Lopes da Cunha and J. Rabassa (eds.), *Festivals and Heritage in Latin America*,
The Latin American Studies Book Series,
https://doi.org/10.1007/978-3-030-67985-9_12

to the history of *tropeira* activity. However, researching the institution and the subject in question, it can be seen that later on other museums, memorials and collections are established in several places in the country. Therefore, this article deals with Brazilian historiography in relation to the subject and the construction of places of memory of the *tropeiro* in Brazil and its implications in relation to the resonances of cultural heritage taking Castro's museum as a reference.

Keywords Museu do Tropeiro · Museum · Muleteers · Heritage · Monument · History of Brazil

1   Introduction

The *Museu do Tropeiro* (Museum of Tropeiro), in activity for more than 40 years, is the main object of a doctoral research in history that aims to reflect on the notions of heritage and memory from the trajectory and consolidation of this museological institution. The English term closest to the term "tropeiro" is "muleteers", just as the expression "mules train" helps to understand the idea of "tropa de mulas". However, considering the specificity of the subject, we chose to use the Portuguese words in the rest of the text. This chapter deals with the reflections built up through the process of investigation, especially with regard to historiography on the *tropeiro* (or muleteers) activity in Brazil and the relationships established with the conception of a monument. It seeks to understand the production of historical knowledge on the subject in order to draw possible parallels with the institution's narrative and the establishment of a place of memory. It is in this way that the text presents the reflections raised so far and proposes the dialogue between memory, history and heritage. It is important to note that the reflections also rely on the experience of the teacher advisor at the Museu Histórico e Pedagógico de Ourinhos (Historical and Pedagogical Museum of Ourinhos) (Cunha 2009, 2014).

2   The Figure of the *Tropeiro* in Brazilian Historiography

The term *tropeiro* (or muleteers) refers to the person who drives a *tropa* (trains), mainly of horses or mules; the one who drives beasts of burden or herds of cattle, such as horses and oxen; the one who trades cattle, finally, a transport entrepreneur. The historian Moacyr Flores wrote that "the dictionary of Eduardo de Farias, 1861 edition, defineds a *tropeiro* as a man who travels with cargo horses" (Flores 1998, p. 19). For Tiago Luís Gil, as explained in his thesis, *Coisas do caminho Tropeiros e seus negócios do Viamão à Sorocaba* (1780-1810) the *tropeiros* on the Viamão-Sorocaba route were not organized as a concise social group, since "they did not have

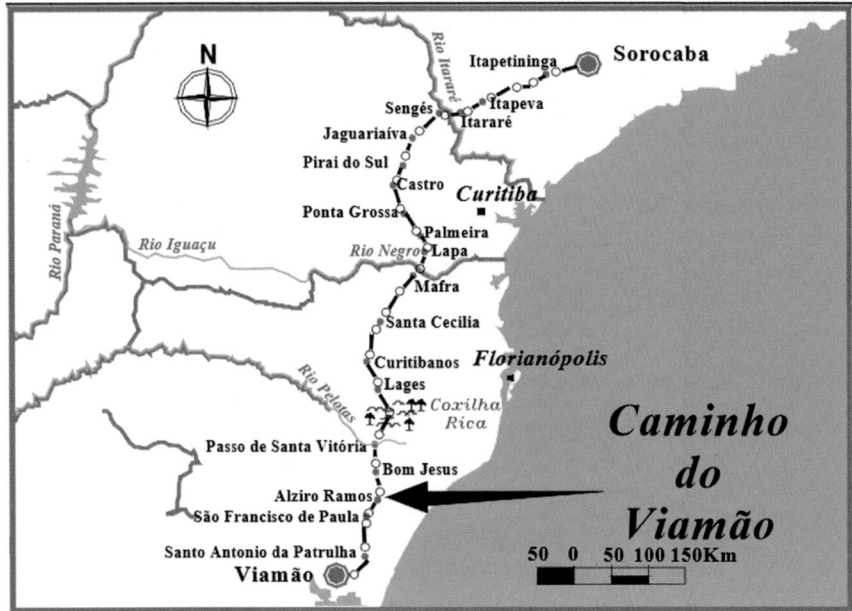

Fig. 1 *Caminho do Viamão* (on a current map)—Organization: Rodrigo Meira Martoni. *Source* Martoni 2005, p. 71

ethnic, political or class identity, even though their social performance was guided by the public image that these animal dealers had" (Gil 2009, p. 51) (Fig. 1).

Some researchers, such as the aforementioned historian Moacyr Flores (1998) and the priest and writer Aluísio de Almeida (1968), believed that stumbling activity can be considered a worldwide phenomenon since animal traction was once the main means for locomotion within a given territory. In Brazil, there is a record of cattle fairs in 1614 in the locality of Capuame, state of Bahia, from which it can be seen that people and animals were moving and these were the object of commercial transactions soon after the Portuguese colonization. However, the axis of research that generated this article is later in time, and differs in space, since it deals specifically with trade of animals in southern Brazil in the eighteenth century.

One of the recurring points in the bibliography is the question of the territorial expansion of the country from this activity that was unfolding and creating the necessary logistic network through the places where it passed.

> Cattle-raising, despite having been a secondary and accessory activity in the colonial period, always being in the background and depending on the other large economic explorations, dare now, export agriculture, especially sugarcane and mining, had an extraordinary role in the exploitation and occupation of vast areas of Brazil today (Holanda 2003, p. 244).

The best known and main route of this trade, called "Caminho do Viamão" (Viamão Pathway), had as its initial term the trip of Cristovão Pereira de Abreu that took place between the years 1731 and 1732. During this first entourage, approximately 3000 mules were driven, thus "inaugurating" a pathway that linked the region of Viamão to the fields of Curitiba (Goulart 1961, p. 37). This route was being opened since 1727 by Francisco de Souza e Faria, under the order of the governor of the captaincy of São Paulo Antônio da Silva Caldeira Pimentel. "The same governor would have created, in February 1732, the Registro de Curirba (Registry of Curitiba), an institution that would control the collection of taxes on animal movements on that new route" (Gil 2009, p. 45). Gil also pointed out that since the early years this route has been widely used, showing the size of the traffic of men and animals.

> A document from the late 18th century estimates that between January 1734 and September 1747, the yield was 42:326$580. Considering that the tax paid at this time, both for horses and mules, was 1$000, we concluded that it passed the equivalent of more than 42.000 animals, over thirteen years and a few months, in an average of approximately 3.200 annually. Between October 1747 and September 1759, when the income was divided in two, half for the Real Fazenda, half for Cristóvão Pereira de Abreu as a result of his deeds, the income of the Real Fazenda was 84:396$810. For this period there is a list of *tropas*, individually listed and described, for the year 1751, which recorded the passage of 9.502 head of cattle (Gil 2009, p. 45).

In 1750, the *Registro de Sorocaba* (Sorocaba Register) was created, as it was there that the biggest animal trade point of the route was developed, including the establishment of an annual fair: the *Feira de Sorocaba* (Sorocaba Fair). Based on these facts and registers, we can see how intense was the traffic of *tropas* and *tropeiros* crossing the regions that today correspond to the state of Rio Grande do Sul, Santa Catarina and Paraná until reaching the state of São Paulo. Sérgio Buarque de Holanda (1975) also stated that it is from 1730 onwards, "with the opening of the road by land that from Curitiba would lead to the Viamão fields and the *Colonia del Sacramento*, that the horse begins to take place in the ordinary rhythm of life in São Paulo" (Holanda 1975, p. 154). In the work *Caminhos e Fronteiras*, there is a passage entitled *Do peão ao tropeiro*, in which the historian wrote about the lack of transport animals and the condition of roads in the region of São Paulo. "The *paulista* (people from São Paulo) used to travel on foot, because to have horses of their own was a luxury" (Holanda 1975, p. 131).

The *Caminho do Viamão*, despite being the main or best known, was not the only route that supplied such commercial activity. There are records of other routes such as the *Caminho da Praia* (Beach Path) and the *Caminho das Missões* (Missions Path), which left and passed through different places in the Southern region. However, when such travellers entered the territory that today belongs to the state of Paraná, all the pathways used to go towards Sorocaba converged to the Campos Gerais region. Reinhard Maack (1981) defined Campos Gerais, geographically, as a region of approximately 19,060 km^2 "used predominantly for intensive cattle raising, (…) from the border with the State of Santa Catarina to the limit with the State of São Paulo" (Maack 1981, p. 256). The Dicionário Histórico e Geográfico dos Campos Gerais (Melo 2019), although based on Maack (1981), adopted a definition that preserves the natural and historical criteria of regional identity and, at the same time, also considers the current configurations of the territory. It can be stated, in accordance with such references, that the region includes municipalities that had primitive vegetation of clean fields and that have a link with the *tropeira* activity.

Thus, settlements that were small parishes and villages like Castro, Príncipe (Lapa), Ponta Grossa, Palmeira, Piraí do Sul and Jaguariaíva have foundations related to the passage of *tropas* and farms of breeding and wintering (where the cattle brought from the South were fattening and recovering vigour to continue their journey). To understand what the landscape was like in this region, and why *tropeiros* preferred these stops, we bring the record of August Sant-Hilaire's impressions in 1820:

> These fields are undeniably one of the most beautiful regions I have travelled since I arrived in America; their lands are less flat and do not become as monotonous as our Beuce plains, but the undulations of the terrain are not as sharp as to limit the horizon. As far as the eye can see, there are extensive pastures; small capons where the valuable and imposing Araucaria stands out, appear here and there in the lowlands, the tone loaded with its foliage contrasting with the light green and lush of the capinzal. From time to time they point rocks on the hillsides, from where a curtain of water is poured that will be lost at the bottom of the valleys (...) (Saint-Hilaire 1995, pp. 15–16)

As far as Brazilian historiography is concerned, we take as a sample for this text the works *Caminhos e Fronteiras* originally published in 1957 and volume 1, volume 2 of the work *História Geral da Civilização Brasileira* published in the same way in 1997, both organized by Sergio Buarque de Holanda, the latter with the participation of other authors. As it has already been pointed out, cattle ranching is remembered as a minor economic activity in the colonial economy; however, it is admitted that "cattle ranching played an important role in the geographical formation of colonial Brazil" (Holanda 2003, p. 255). This latter book also brings that:

> Cattle cars and animal train ensured economic development, the profits produced by livestock farming, unlike mining and commercial agriculture, were incorporated into the country, contributing to its material progress, despite the little that the metropolis has done to assist this economic activity (Holanda 2003, p. 255)

To bring a contemporary look at the subject, we resort to the thesis already cited by Tiago Luis Gil (2009), which has as its main objective the understanding of economics from the study of how credit was practiced in the dynamics of subjects' personal relationships. For the author, the organization in front of the Viamão-Sorocaba trade route can be called a "corporation of captains".

> It was a group with an analogy to craft corporations, but not one, formally. They maintained a social control among themselves, which was allowed by the way they themselves circulated and circulated information. In terms of relationships, they placed themselves above community relations, but also within them, participating in different layers of relationships and profiting from them. In the end, they could control their own movement of men, women and resources within that vast territory. But for such a corporation to be made daily by the captains, there had to be other fixed points: Catholicism, the monarchy, notions of equity, corporate society and family (Gil 2009, p. 257).

Gil demonstrated through his research that not everyone who worked in the animal business was called a *tropeiro*. The owners of the *tropas* often used military patents, and there were others who worked in other people's *tropas*, called pawns or foremen.

> My observation is that the animal business was not manipulated by a specialised group or controlled by some monopoly. It was an open market possibility, available not only to prominent members of a local elite, but also to a large number of people who were able to obtain some resources or a loan (Gil 2009, p. 56)

The author makes no reference to the naturalist August Saint-Hilaire, but he made a similar observation in the same register cited above:

> One should not think, however, that the inhabitants of the Campos Gerais will always remain in their region. Men of all classes, workers, farmers, at the moment they earn some money, leave for the south, where they buy brave donkeys to resell them on their own land or in Sorocaba. (Saint-Hilaire 1995, p. 19).

Local historiography (José Pedro Novaes (1972); Oney Barbosa Borba, 1986), points out that the Iapó river, due to its characteristic of becoming flooded, forced the *tropeiros* that went there to wait and camp. It is in this scenario, still in the colonial period, that the *Pouso do Iapó* was born, which in 1774 was constituted as a parish of *Sant'Ana do Iapó*, becoming then one of the first administrative organizations in the Campos Gerais region. In the mid 1970s, in this same place now called Castro (since its elevation to the village in 1789), Professor Judith Carneiro de Mello wished to preserve and divulge the history of the city through the creation of a museum. Based on local and regional historiography, she headed the foundation of the Museum of Tropeiro.

3   The Uses of History: The Dialogue Between Memory and Heritage

To understand the trajectory of the Museum of Tropeiro and other spaces designed to remind these social actors, their activities and practices, it is necessary to ponder on the concepts of heritage and memory. One can list as references the historians Françoise Choay (2017) and Sandra Pelegrini (2009), the archaeologist Pedro Paulo Funari (2009), the historians Pierre Nora (2008) and Jacques Le Goff (2013) and the anthropologists José Reginaldo Gonçalves (2005) and Nestor Canclini (1994). Repeatedly, according to common sense, the museum is seen as a static, old and dusty place, possessing the function of guarding "old things" that no longer have any use or a place of the intellectual and economic elite for safeguarding the history of heroes and rulers.

Jacques Le Goff (2013) argued that the term "ancient" oscillates between wisdom and senility, that is, synonymous with old as pejorative. However, with the Renaissance, the "old" gains a new perspective, thanks to the appreciation of Greek-Roman Antiquity. It was during this period that movements such as the Renaissance and

Humanism relied on each other to underpin the modernity of the sixteenth century. However, it is only with the Enlightenment that the idea of a cyclical time will be replaced by the idea of a linear progress that significantly privileges the modern. In relation to the past and the present, the French historian shows that the cognition of time varies according to society and time. At the beginning of the twentieth century, the crisis of progress motivates new attitudes towards the past, present and future. In the post-war period, the link with the past begins to acquire reactionary configurations, since there is anguish in the face of atomic weapons and euphoria in the face of scientific progress. In this sense, one turns to the past with nostalgia and to the future with fear or hope. The acceleration of history has also led the peoples of industrialized countries to nostalgically connect to the past. "Hence the retro fashion, the taste for history, archaeology, enthusiasm for photography and prestige of the notion of heritage" (Le Goff 2013, p. 210).

The classic text entitled Document/Monument also helps in this reflection, since Le Goff proposed to think about how the question of memory arises in the human sciences, especially in history and anthropology. The author warns that disorders in memory are not only disturbances of the individual, but have reflections on memory and collective identity. In this sense, the first point explored by him is that of ethnic memory, which solidifies the collective memory of peoples without writing, the one that underlies the existence of ethnic groups or families—the myths of origin. It is important to realize that it is not characterized as a memory word for word. In this way, the author deals with the variations that exist in myths precisely because the process of memorizing these unwritten societies has more freedom and creative possibilities. He also explained that, in societies without writing, the collective memory has three main objectives: the knowledge of the collective age of the group (myths of origin), the prestige of the dominant families (genealogy) and technical knowledge (religious magic). With the appearance of writing, there is a significant transformation in collective memory (Le Goff 2013, p. 394).

Writing allows collective memory to record the origin and commemoration of its events. This is how commemorative monuments such as stelae or obelisks with inscriptions and images appear. In Egypt, funerary monuments stand out, and there are also legal stelae, such as the code of Hammurabi in Mesopotamia. These records gain visibility with the Greeks and Romans and reach the paper support. Françoise Choay also warned of the affective nature of monuments in relation to memory, since the monument acts on the memory making it "vibrate as if it were present" (Choay 2017, p. 18).

> The monument assures, calms, reassures by conjuring up the being of time. It constitutes a guarantee of the origins and dispels the restlessness generated by the uncertainty of the beginnings. Challenging entropy, the dissolving action that time exerts on all natural and artificial things, it tries to combat the anguish of death and annihilation (Choay 2017, p. 18)

The creation of archives, monuments and museums is essential in this process of memory storage and conservation. The myth of origin of museums contributes to the understanding of these institutions. The term Mouseion means temple of the muses, which refers to the nine daughters of Mnemosyne, goddess of memory, and

Zeus, the supreme god of Olympus. It is interesting to note that this myth suggests the union between power and memory, and so the museum is a space where power and memory walk side by side. Le Goff recorded that memory is first perceived as a gift, for the goddess Mnemosyne revealed to the poet the secrets of the past and introduced him to the mysteries of the beyond. By this bias memory is placed outside of time, and there is a separation between memory and history, "memory can lead history or distance itself from it" (Le Goff 2013, p. 401).

4 The Tropeiros and Their Places of Memory

The places of memory are, first and foremost, remains. The extreme form where a commemorative conscience subsists in a history that calls it, because it ignores it (...) Museums, archives, cemeteries and collections, parties, anniversaries, treaties, verbal processes, monuments, sanctuaries, associations, are landmarks witnessing another era, the illusions of eternity (...)....) Places of memory are born and live from the feeling that there is no spontaneous memory, that it is necessary to create archives, that it is necessary to maintain anniversaries, to organize celebrations, to pronounce funeral praises, to notarize minutes, because these actions are not natural (Nora 1993, pp. 12–13).

It is from this understanding that we propose the problematization of the places and the actions elaborated with the intention of ritualizing and celebrating the history of a mercantile activity that was born in the period of Colonial Brazil, developed and structured during the eighteenth century, and that still resisted in some places in the country until the decade of 1950, being gradually replaced by railroads and definitely contained with the implantation of roads.

With the advance of such modernization of transportation and the closing of the tropeira activity, the eagerness to preserve such memory arises, which drives the perpetuation of the narrative and the objects tangent to it. In this context, with the support and guidance of the historian from the state of Paraná, Newton Carneiro (historian, university professor and collector, a specialist in art history and iconography of Paraná), teacher Judith Carneiro de Mello organized and instituted a historical museum in the city of Castro, the Museum of Tropeiro. Judith was born in the city of Castro on April 11, 1923, daughter of Maria da Conceição Bueno Barbosa Carneiro de Mello and Vespasiano Carneiro de Mello, a businessman and politician from Paraná. She worked as a teacher and school principal and retired in 1983, after thirty years dedicated to Education. The project was materialized with resources from Castro's City Hall, and was created on December 1, 1976 and inaugurated the following month, in January 1977.

On the twenty-first day of January, one thousand nine hundred and seventy-seven, the hundred and twentieth year of the elevation of the city of Castro to the category of city, the fiftieth year of the Independence of Brazil and the year of the Proclamation of the Republic was officially inaugurated, in the presence of the distinguished authorities and the public undersigned the Museum of Tropeiro, with the objective of evoking that historical figure of Colonial Brazil, from the 18th and 19th centuries, who contributed so much to the formation of Santana do

Iapó, today the city of Castro, as well as many other cities in the Provinces, today the States of Rio Grande do Sul, Santa Catarina, Paraná and São Paulo. (Museu do Tropeiro 1977, p. 1)

About the intention to create a museum, it is possible to observe in the speech of Mrs. Judith the purpose of creating a place of memory:

I remember that Dr. Newton Carneiro, a great historian from Paraná, was the one who aroused my desire to know Castro's history. Because, until then, I didn't give much importance, because I didn't hear about Castro's history. When Dr. Newton came with me to the house where we would install the Castro's Museum, he said: Look, Judith, I'm sorry to make a guess like that, my arrogance, but I think here, we'd be making a museum of the tropeiro, which is Castro's origin. So I thought and answered "Dr. Newton, you are here to guide me, because I intended to make a museum, because I can see that all Castro's things, Castro's customs are evaporating. (Associação de Amigos do Museu do Tropeiro 2013, p. 25).

The idea of a museum seems to be tied to the possibility of preserving "things that are evaporating". It is possible to understand part of the institutional training discourse through documents such as the opening minutes, the founder's discourse, agendas and notebooks that are being researched and that constitute the list of possible sources for the research. For the moment, it is pointed out what the bulletin commemorating the third anniversary of the museum published and distributed in 1980 brings.

Despisers of death, proud to be called: brave, worthy flagship lineage of independent feudal knights, individualists, carrying the law at the point of a sword or a machete, but always with a minimum of standards of conduct and dogmas of honorable people (Almeida in Museu do Tropeiro 1980)

The quote of the writer and researcher Aluísio de Almeida on the back cover of the document is an exaltation to the figure of the *tropeiro*. There is no criticism or questioning, it only highlights what the author would judge as qualities. According to the documentation consulted, the institution was created with a speech of homage; however, it is also significant to observe the research bias that appears since its foundation.

The Museum of Tropeiro, during the years 1977, 1978 and 1979, fulfilled its cultural mission by acquiring, studying, preserving and showing. The museum began its informative didactic objective of showing the artistic and historical culture of the region, aiming besides portraying the life of the tropeiro, to present historical documents, old weights and measures, sacred pieces and also a comparative picture of the present days. Not only did it expose but it also facilitated the researchers their knowledge (Museu do Tropeiro 1980)

The collection was mainly constituted by donation, but there is also registration of purchases by the City Hall and by Mrs. Judith. The material culture related to this theme is quite diverse. The handling of animals is composed of several very specific objects: *freios*, saddles, *pelegos*, stirrups, *coxonilhos*, whips, spurs; objects of cargo, such as *bruacas, canastras, cangalhas, jacás*; objects of clothing, such as hats, boots, *ponchos*; objects of use in the landings, benches, cauldrons, *chocolateiras*, cutlery; objects of personal use, *garruchas*, knives, daggers, canes; among other artifacts that are directly related to work, travel and the trade of animals

Besides the composition of the three-dimensional collection, the institution also turned to the organization of a documentation centre and a specialized library, proposing, encouraging and enabling the construction of knowledge through research. The 1970s is considered a milestone in the history of museology and museums around the world since there is the strengthening of a movement that proposes new perspectives and new approaches on these institutions. The so-called New Museology brings, among other aspects, discussions about the social function of museums. Thus, it is disturbing to think that despite being at first a traditional institution, created to honour a certain "hero", the text of the bulletin also says that "the Museum of Tropeiro is not a depository of old things, but a house where the past is lived and even though it lacks many requirements to be a model museum, it has sought to be active" (Museu do Tropeiro 1980).

According to the documents researched in the administrative archive of the institution and also a survey in the Cadastro Nacional de Museus (national museum register) made available by IBRAM, the Brazilian Institute of Museums, the Castro museum is the first one created exclusively for the theme of the history of *tropeiros*. However, we find the register of eight other organizations according to the Table 1:

Table 1 Museums, collections or memorials dedicated to the *tropeiro* in Brazil

Name	Foundation	Status	City and State	Administration
Museu do Tropeiro	1977	Opened	Castro-Paraná	Public
Centro de Estudos do Tropeirismo;/Casarão de Brigadeiro Tobias	1981	Opened	Sorocaba-Sãom Paulo	Public
Museu do Tropeiro	2003	Opened	Itabira (Ipoema)-Minas Gerais	Public
Museu do Tropeiro—Seccional Casa de Sinhara	2004	Opened	Castro-Paraná	Public
Acervo Tropeiro de Tupã*	2004	Opened	Tupã-São Paulo	Private
Museu do Tropeiro José Salomão Fadlalah	2011	Closed	Ibatiba-Espirito Santo	Public
Museu Tropeiro Velho*	–	Opened	Chapecó-Santa Catarina	Private
Acervo Tropeiro de Carambeí	–	Closed	Carambeí-Paraná	Private
Museu do Tropeiro e Centro de Artesanato Aloísio Magalhães	–	Opened	Lapa-Paraná	Public
Museu dos Tropeiros	–	Closed	Entre Rios de Minas-Minas Gerais	Private

Source Brazilian Institute of Museums (IBRAM). National Register of Museums. Available at: http://museus.cultura.gov.br/. Accessed on December 04, 2019

Table 2 Tropeiras cities in the southern region of Brazil

State	No of cities	Tropeiras Cities	%
Rio Grande do Sul	497	49	9,86
Santa Catarina	295	48	16,27
Paraná	399	29	7,27
Total	1191	126	10,58

Source Alves, L. A.; Alves, S. M. S. 2018. Cidades tropeiras: região sul do Brasil. Porto Alegre: Evagraf, p.13

In addition to these records found on the online platform of the *Cadastro Nacional de Museus*, it is believed that there are other institutionalized spaces or private collections. From the cutout that mainly covers the southern states and the state of São Paulo, many localities recognize its connection with the tropeira activity. There is also a 2018 publication entitled *Cidades Tropeiras: região sul do Brasil* by Luiz Antônio Alves and Sandra Maria Schmith Alves on research conducted between the years 2014 and 2016 on official websites of city halls in search of those who cited tropeira activity in their pages. The authors state that of the 1191 municipalities in the southern region of the country 126 make this reference (Table 2).

As researchers of the subject, the present authors would like to warn that the pages provide little information and that some cities admittedly involved with the activity do not present any communication. However, it can be suggested that the absence is not purely constituted of negligence or omission, but of a lack of identification or selection of diverse narratives. According to what has been pointed out, the anthropologist José Reginaldo Gonçalves proposed a reflection on the categories resonance, materiality and subjectivity in order to understand the heritage discourses. For the author, the analyses on cultural heritage have emphasized its "built" or "invented" character. "Each nation, group, family, at last each institution would build in the present its heritage, with the purpose of articulating and expressing its identity and its memory". (Gonçalves 2005, p. 214).

Gonçalves (2005) wrote that it is necessary to analyse the cases in which state policies do not find recognition among the population, that is, "a patrimony does not depend only on the political will and decision of a state agency. Nor does it depend exclusively on the conscious and deliberate activity of individuals" (Gonçalves 2005, p. 214). It is based on the concept that museums and the theme proposed by them are historical and historiographic constructions, as well as the selection of certain objects and documents characterized as heritage is subject to choices and perspectives.

Analysing the case of the Museum of Tropeiro, it is possible to affirm that the collection shown in a long-term exhibition intends a narrative that exalts the elite of the Campos Gerais region through the imposing furniture, silver objects and imported china, an elite that dominated the political and social space. However, to tell the story of the daily life of the tropeira activity, it is necessary to talk about all the actors. There are other objects that also make up the collection and concern the daily life of a rural Brazil, which ends up bringing together and reaching the most different subjects. "The museum and any patrimonial politics must treat the objects, the trades and the customs in such a way that, more than exhibiting them, they make intelligible the relations between them, propose hypotheses about what they mean for the people who see and evoke them today" (Canclini 1994, p. 113).

The doctoral research that is underway will deepen these reflections through the analysis of sources such as the diaries produced by the founder and the employees of the Museum of Tropeiro over 30 years (1977-2007),[1] letters received and sent, newsletters from the institution, public counting through the visitors' book in order to quantify and classify their origin, as well as news published in local and regional newspapers and periodicals. It is believed that the analysis of the sources and the understanding of the institution's trajectory, together with the interlocution with anthropology, will make it possible to answer the gaps about the dialogue between memory and history in the face of the construction of a place of memory and of a narrative that institutionalizes and validates as cultural heritage.

References

Almeida A (1968) O Tropeirismo e a Feira de Sorocaba. Luzes, Sorocaba

Associação de Amigos do Museu do Tropeiro (2013) Judith. Gráfica Kugler, Castro

Alves LA, Alves SMS (2018) Cidades tropeiras: região sul do Brasil. Evagraf, Porto Alegre, p 13

Canclini NG (1994) O patrimônio cultural e a construção imaginária do nacional. Revista do Patrimônio Histórico e Artístico Nacional 23:95–115

Choay F (2017) A Alegoria do Patrimônio, 6th edn. Estação Liberdade: UNESP, São Paulo

Cunha FL (2009) Histórias e memórias de um "museu local": o museu histórico e pedagógico de ourinhos, Patrimônio e memória, UNESP – FCLAs – CEDAP, vol 4, no 2, pp 163–197

Cunha FL (2014) Memórias dos Trilhos: Um guia prático sobre revitalização de museus, 1st edn. UNESP-Campus Experimental de Ourinhos, Ourinhos

Flores M (1998) Tropeirismo no Brasil. Nova Dimensão, Porto Alegre

Funari PP, Pelegrini SCA (2009) Patrimônio Histórico e cultural. 2nd edn. Zahar, Rio de Janeiro

Gil TL (2009) Coisas do Caminho: Tropeiros e seus negócios do Viamão à Sorocaba (1780–1810). Unpublished doctoral thesis in Social, Programa de Pós-graduação em História Social, Universidade Federal do Rio de Janeiro, Rio de Janeiro

Goulart JA (1961) Tropas e Tropeiros na formação do Brasil. Conquista, Rio de Janeiro

Gonçalves JRS (2005) Antropologia dos objetos: coleções, museus e patrimônios. São Paulo, n, Revista Brasileira de Informação Bibliográfica em Ciências Sociais BiB, p 60

Holanda SB (1975) Caminhos e Fronteiras. J. Olympio, Rio de Janeiro

Holanda SB (2003) História Geral da Civilização Brasileira. tomo I, vol 2. Bertrand Brasil, Rio de Janeiro

Le Goff J (2013) História e Memória. Editora da Unicamp, Campinas

Maack R (1981) Geografia física do Estado do Paraná. Rio de Janeiro, Livraria José Olympio Ed, 442 p

Martoni RM (2005) Caminhos redescobertos: o potencial turístico das rotas doSul. 183 p. Unpublished Master Thesis Geografia, Environment and Development, Universidade Estadual de Londrina, Londrina

Melo MS (2019) Dicionário Histórico e Geográfico dos Campos Gerais. https://www2.uepg.br/dicion/. Accessed 12 Dec 2019

---

[1] Judith Carneiro de Mello was in charge of the institution until the year of her death in 2007.

Museu do Tropeiro. Castro. Ata de inauguração realizada no dia 21 de janeiro de 1977. Livro 1, 200, pp 1–2

Museu do Tropeiro. Castro. Boletim Informativo.1980

Nora P (1993) Entre memória e história: a problemática dos lugares. Projeto História, São Paulo, no 10, pp 7–28

Nora P (2008) Pierre Nora en Les lieux de mémoire. Trilce, Montevideo

Saint-Hilaire A (1995) Viagem pela Comarca de Curitiba. Fundação Cultural, Curitiba

Archaeological Heritage and Tourism

Illicit Trafficking in Cultural Assets: A Genealogy of the Concept and Actions in Contemporary Brazil

Rodrigo Christofoletti

Abstract A significant part of cultural heritage has been taken by a multimillionaire trafficking system, and documents from international organizations estimate that art and heritage trafficking tops the list of the biggest illicit actions in the world, just behind drugs, weapons and human trafficking, which created a well-articulated systemic grid that indicates an exponential growth trafficking pattern. Such a mesh feeds a fairly complex system. Some examples of this mesh are private collectors, museums, monuments, religious sites, archeological/paleontological sites and other private preservation spaces; illegal excavations (including underwater excavations); theft of artifacts and works of art during armed conflicts and military occupations; illegal downloading of miscellaneous properties; production, exchange and use of forged documentation; even trafficking of cultural goods, authentic or counterfeit. This whole range of actions linked to trafficking has been fought in recent decades as the life of this set of goods is increasingly in danger. This article aims to address this type of illicit trafficking, suggesting that an international art trafficking route has in Brazil one of the least studied but no less important capillary points.

Keywords Cultural Heritage · Illicit trafficking · Archeological · Paleontologica · Museums · Monuments

It is important, from the Brazilian point of view, to maintain the balance of action in the international community in all aspects of the process aimed at the protection of cultural property, participating in the negotiation processes and implementing the obligations and duties of the conventions to which it is a signatory. A country traditionally engaged in articulations aimed at distention, it is also responsible for maintaining a coherent multilateral action, participating in the processes and decisions related to the protection of cultural heritage in case of illicit trafficking of cultural goods. Therefore, it is important to register the process that has made Brazil one of the most attentive signatories to the traffic of works and goods, as described below.

(Christofolletti 2017)

R. Christofoletti (✉)
Federal University of Juiz de Fora—UFJF, Juiz de Fora, Brazil
e-mail: r.christofoletti@uol.com.br

For about a decade now I have been coming through texts and actions in entities representing the preservation of cultural heritage discussing the need to deepen our understanding of the real meaning of the gears of illicit trafficking of cultural goods around the planet, and especially in Brazil. The epigraph that introduces this text serves as a parameter for us to introduce this discussion. The cultural heritage has been squandered by a multi-million dollar trafficking system, and documents from international organizations estimate that art and heritage trafficking is at the top of the list of the world's largest traffickers, behind only the smuggling of drugs, weapons and human beings (UNESCO 2011). Covering numerous activities, ranging from the export of cultural goods by their legitimate owners without the necessary authorization to specialized trade in stolen objects, and the appropriation and commercialization of works of art unknown to the authorities, this modality has caused concern to the States, both the dispossessed and the recipients, because both are directly affected by theft and illicit trafficking of cultural heritage.

Although the trafficking of cultural goods has gained space in academic texts, alongside other forms of trafficking, there is no exact legal definition of the acts that this term penalizes. The broad definition of trafficking can be considered as any movement, transport, import, export, maintenance or trade of cultural goods carried out in violation of the rules governing the possession or circulation of such goods or their status. But, after all, what does illicit trafficking of cultural goods consist of? It is the flow, commercialization and use, in an illicit way, of goods with a cultural dimension, of different natures: art objects and paleo-archeological artifacts, objects of art and religious cult, documents coming from official archives, antiquities in general, manuscripts and rare bibliographical works, phonographic and ethnological material, that is, a wide range of possibilities and symbols.

The systemic and very well-articulated machinery that characterizes the scheme of trafficking in cultural goods ranges from the theft of museums, monuments, religious sites, archeological/paleontological and other private and public preservation spaces; illicit excavations (including underwater); subtraction of artifacts and works of art during armed conflicts and military occupations; illicit export and import of artifacts; illegal transfer of ownership of various cultural goods; production, exchange and use of falsified documentation; even trafficking in authentic or falsified cultural goods themselves. All this list of actions has been fought in the last decades, a factor that helps to widen the visibility of cultural goods in danger around the planet.

Combating attacks on archeological, historical and artistic riches requires international cooperation, both in preventing infringements and in ensuring the restitution of the property removed. This work aims to address this type of illicit trafficking, suggesting that the international route of trafficking in works of art has in our country one of the least studied capillarity points. The understanding of this missive helps to understand how the illicit trafficking of cultural goods and works of art is today the third most important in financial volume in the world, moving more than 6 billion dollars in the last decade, according to the FBI, Interpol and UNESCO. In this sense, the policy of repatriation of trafficked goods is a challenge to contemporary States, a factor by which the study of this theme becomes relevant.

This text is in dialogue with another reflection produced by me, present in the text: *The illicit trafficking of cultural goods and repatriation as historical reparation* which was published in (Christofoletti 2017). It also dialogues with journalistic reports compiled over the last decade in blogs, websites, newspapers of great circulation, and what is most contemporary in the literature and historiography on the subject. (Alford 1994; ASkerud 2019; Feliciano 1997; Frammolino 2011; Horta 1996; Johnston 1993 La Follette 2013a; Merryman 2008; Miles 2008; Nicholas 1996; Todeschini 2007; Waxman 2011).

On a global scale, common procedures have been sought to combat trafficking in works of art and cultural goods. The 1970 UNESCO Universal Declaration on Cultural Diversity already proposed the elaboration and application of policies and strategies for the preservation and enhancement of cultural heritage and the fight against illicit trafficking of cultural goods and services. Underwritten by more than 80 countries, the declaration has not been complied with by some of the countries considered as target markets for illicitly trafficked works of art. Also within the framework of the United Nations, the Convention of the International Institute for the Unification of Private Law (UNIDROIT) was signed in 1995, which deals with the recovery of stolen cultural works. In addition to Interpol and UNESCO, World Customs Organization (WCO), International Council of Museums (ICOM) and Federal Bureau of Investigation (FBI) are responsible for combating trafficking.

The 1970 Convention drawn up by UNESCO lists measures aimed at prohibiting the import, export and illegal transfer of cultural goods. The purpose of this convention, therefore, is to protect the world's cultural heritage by means of rules to be implemented by the 89 countries that signed the treaty. Among the main measures suggested by the UNESCO Convention, the following should be highlighted: the creation of appropriate national legislation to combat illicit trafficking; the establishment and implementation of a national inventory system, with the purpose of listing all cultural works; the requirement of an export certificate, which must accompany any type of cultural good eventually exported; the creation of a code of ethics for art collectors and dealers; the implementation of educational programs to provide respect for cultural heritage and rules to ensure that any interested party can denounce the disappearance of cultural goods.

It should be stressed that the UNESCO Convention also requires the imposition of appropriate penalties and sanctions in order to curb the practice of illicit trafficking in cultural goods. Therefore, the fight against illicit trafficking in cultural heritage is characterized by its mixed legal nature, regulated simultaneously by Administrative Law, Criminal Procedural Law and Criminal Law. It is recommended, in relation to this topic, to consult the laws listed as follows: Decree-Law No. 25/37 (it organizes the protection of the national historical and artistic heritage); Law No. 3,924/61 (it provides for archeological and prehistoric monuments); Law No. 4,845/65 (it prohibits the departure abroad of works of art produced in the country until the end of the monarchic period); Law No. 5,471/68 (it provides for the export of old books and Brazilian bibliographic collections); Decree-Law No. 72,312/73 (arising from the UNESCO Convention in 1970) and Ordinance 262/IPHAN.

In reality, it is difficult to pinpoint the depth of the depredation of this form of trafficking. One of the reasons that makes it difficult to quantify the illicit is that the value of cultural goods is not the same in the country of origin and destination. On the other hand, part of the theft is not reported, since the purchases were made without tax returns. It is also impossible to quantify the damage caused by objects subtracted from clandestine archeological excavations. According to Interpol, the countries most affected by thefts are Germany, France, Italy, England and Russia. Interesting reversal, if we think it was these same nations that over the past three centuries have become the major responsible for the misappropriation of cultural goods and works of art around the world. See their museums. It is estimated that private individuals are the target of the greatest number of robberies, followed by museums, archeological sites and places of worship. Official sources claim that paintings, sculptures, statues and religious objects are among the most trafficked artifacts.

Cultural heritage and its interface with international relations are part of an expanding field of study. There is a positive agenda in the case of heritage preservation and it is in this context that the approach between heritage and international relations is inserted. At the end of the twentieth century, classic themes worked on in the field of internationalism, such as the polarity of the international system, and balance of power, security, among others, gave way to other theoretical lineages, which multiplied their productions on themes more reactive to contemporary reality, such as human rights, the environment and diversity. In this scenario, cultural and heritage preservation issues came to the fore due to concerns about international security.

To understand the interface that brings cultural heritage and international relations closer together, the expression "soft power", by Joseph Nye Jr. (1990), helps us to search for paths that are still little traced. Soft power is the ability to influence others to do what you want by attraction rather than coercion. In international politics, power is considered a means and an end by which a relationship of domination of one party over the other develops, which guarantees one party the power to determine the behavior of others in pursuit of their interests. This definition serves as a *leitmotiv* for understanding the role of preserving cultural heritage in international relationships.

Within the multilateral framework, efforts to institutionalize new preservation practices have been made on a larger scale since the mid-1970s, emanating mainly from UNESCO. Some documents have guided these efforts, such as the *Convention on the Prohibition of the Import, Export and Transfer of Illicit Property of Cultural Property* (1970), the Convention for the *Protection of the World Cultural and Natural Heritage* (1972), *Universal Declaration on Cultural Diversity* (2001), the Convention on the *Safeguarding of the Intangible Cultural Heritage* (2003) and the *Convention on the Protection and Promotion of the Diversity of Cultural Expressions* (2005). The large-scale use of these legal preservation mechanisms has potentialized the creation of a trend: cultural heritage as the key to new socio-cultural and diplomatic approaches that have been constantly updated by scholars.

The extension of related themes to the universe of cultural heritage has made it possible to understand the protagonist dimension of the immateriality of cultural assets; the presence of other actors in the production and management of heritage; the increase in the modalities of tourism (sustainable or predatory) in human heritage; the intensification of comparative studies between state parties to international organizations; the criteria used for the selection, reception, adherence and safeguarding of policies on cultural assets; the expansion of studies on heritage brought about by waves of immigration, as well as the increasingly multifaceted action of issues related to cultural diplomacy.

International organizations began to see their assets more broadly, placing them on the agendas that make up global governance. Whether it is related to the idea of sustainability, the fight against extremism, or policies around access to citizenship and tradition, the cultural heritage has come to have a relevant participation, with an advance in the presence of preservation bodies at international policy negotiation tables. However, at a time when we are witnessing the worst crisis of cultural heritage on a global scale since the Second World War, with the crimes perpetrated by fundamentalist radicals, reputed as war crimes, two other examples cause astonishment and discomfort.

The large-scale use of these legal preservation mechanisms has potentialized the creation of a trend: cultural heritage as the key to the new socio-cultural and diplomatic approaches that have been deserving of scholars, updating and deepening. This academic care is necessary at a time when we are witnessing a sui generis order: we are witnessing, perhaps, the worst crisis of cultural heritage on a global scale since the Second World War, with the crimes perpetrated by the radicals of the Islamic State considered as war crimes. But radicalism is not the only problem facing heritage on a global scale. Other examples cause amazement and commotion: the material deterioration of heritage around the globe; the low sensitivity of some states to safeguarding their cultural assets; the discharge of the obligations of some states parties, as members of UNESCO, who have disconnected from the institution, causing huge breaches in their coffers, following the example of the United States and Israel; and, mainly, the smuggling of artifacts, which has consolidated itself into a big business, being the third largest illicit activity in financial movement in the world, being behind only the traffic of drugs and weapons, moving more than 6 billion dollars a year, according to official sources. These examples intensify the challenge and the practice of the process of repatriation of these artifacts, although such actions have gained viability in the last decade, as iceberg tips in the increasingly plural dynamics that the world's wealth is facing.

To understand the pluralized interface that brings cultural heritage and international relations closer together, the expression "soft power", coined in the early 1990s in the book *Bound to lead: the changing nature of American power*, by Joseph Nye Jr. (1990), has come to be used by academics and specialized media, mainly in the last decades. In Nye Jr.'s concept, soft power is the ability to influence others to do what you want by attraction instead of coercion. Coercive power would be military ostentation and economic sanctions, classified as raw power, while cultural, ideological and political identity would make up soft power. This concept encompasses and problematizes the multiplicity of topics on the contemporary international agenda, focusing on one of its less discussed elements: the universe of international cultural heritage and the relationship between actors and preservationist actions in the globalized world.

In international politics, power is considered a means and an end by which a relationship of domination of one party over the other develops, which guarantees one party the power to determine the behavior of others in the pursuit of their interests. The definition serves as a leitmotiv to broaden our understanding. By its very nature, "soft power" is a relative and intangible concept, inherently difficult to quantify. The relational nature of soft power gives rise to a substantially complex comparative plan, in which transnational comparisons become complicated and difficult. What is loved in one country, in another, may represent repulsion. Sometimes fought by articulists who see soft power only as a powerful current of maintenance of the American status quo, the concept needs to be understood in its entirety and cannot be understood in a naive way.

At the same time, there is a long list of human enterprises that comes to mind first when modern diplomatic concerns are considered: economy, military affairs, crime, health, environment, terrorism and so on. Even within the list of zones established for international cultural heritage, the generic theme of heritage hardly gets any bigger. The importance of its knowledge is obliterated by other issues of greater general interest, such as contemporary humanitarian crises (refuge, civil wars and field experiences), new trends in foreign policy, contemporary issues of global politics, negotiations or conflicts, among other topics considered more burning. This means that cultural heritage still seeks a more forceful mention in the literature of diplomacy.

Faced with a scenario interconnected by the transmission of information, the concept of Cultural Diplomacy, arising directly from the connection between international relations and cultural heritage, is presented as one of the domains to be discussed. Because of this "thematic novelty", there are still few systematized studies in the field of cultural diplomacy, and currently there is no general consensus regarding its definition. However, what is perceived is that this domain has acted in the understanding of varied themes, functioning as ambassadors of these new world demands. From this, other objects of study, not yet incorporated by cultural diplomacy, are derived: major sports events, such as the World Cup, the Summer and Winter Games and the Olympics; football as the mark of an increasingly globalized soft power; the great artistic and musical festivals around the planet and those of lesser expression, given that they are regional, because they often explain the identity of peoples virtually unknown to the *mainstream*; languages and their frontiers; the dynamics of hierarchization of themes and criteria consecrated by UNESCO advisory bodies; the increasing presence of themes that address the "Africanities", "Asianities", "Latinities" and "Orientalisms" (so little explored by our researchers, given the hegemony of the Europeanist/American vision); among others.

Certainly, the protection of cultural goods on a global level can be considered an intrinsic contribution to human well-being. But in addition to its inherent value for present and future generations, the heritage can also mean an important instrumental contribution to sustainable development in all its various dimensions. With the appropriation of cultural heritage for commercial and political purposes within the economies of all parts of the globe, heritage conservation now plays an important role in cultural diplomacy, elevating its status from a mere diplomatic strategy of good neighborly relations to an elaborate soft power tactic in different countries around the globe. The analyses of heritage governance at the beginning of this century have focused primarily on intergovernmental bodies, to the detriment of a critical reading of the role of nation-states and of paradiplomacia itself, which continue to play key roles in international heritage conservation governance.

In the context of the list of World Heritage managed by UNESCO, the increasingly frequent rotation of representatives of States Parties from various parts of the world (not only from consecrated regions but also economically developed ones) and the capillarization of the map of heritage granted with the seal of "World Heritage" around the world are, in the same way, significant examples of the expansion of soft power as an instrument of power, although for some a more balanced and representative World Heritage list seems to be a mirage as long as the essentials of the classification processes depend fundamentally on the role of national states and as long as World Heritage is excessively attached to an image of symbolic distinction—an important resource of places that seek to become more competitive and media.

These are just a few examples of the areas that international relations have been looking into. As the new century nudges itself, the radiography of these power relations reveals new actors and spaces. Cultural heritage has become an increasingly important contributor to multilateral dialogues and, as such, is part of the broadening of actions in the field of international relations. In this sense, archeological sites, museums, cultural spaces, international preservation organizations, national states, actors of paradiplomacia, expressions of tradition, experience and ways of doing, the dichotomy between inflation and destruction of heritage, among other elements, have become protagonists of this cartography that has been constantly transformed. Understanding the mechanisms of understanding this thematic expansion will favor the establishment of new values of heritage, nationally and internationally.

Therefore, three factors help to consolidate the assertions listed above. The first concerns the increasing incorporation of cultural heritage into other areas of international discourse. Recently, international organizations have come to see heritage more broadly, taking it as part of the discourses and agendas that make up contemporary global governance. Whether it is related to the idea of sustainability, the fight against extremism, or policies around access to citizenship and tradition, cultural heritage has become much more visible and a relevant participation, and there has been an advance in the presence of preservation bodies at international policy negotiation tables as never before seen. The second refers to the pluralization of heritage narratives. There are increasingly challenging expressions of the Eurocentrist/US legacy in the conduct of preservation policies considered homogenizing the preservation of world heritage. Throughout Asia, Latin America, Africa and the Middle East, criticism of the hegemony of these actions has given rise to distinct and emancipatory approaches to curatorship and preservation. The third alludes to the growing economic and political power that countries with preservationist agendas enjoy on the international stage. Unlike the panorama of five decades ago, when only the countries holding economic power dictated the rules of preservation and what could be understood as an exceptional universal value, the current scenario is different. All of this supports the proposition that soft power through culture, exchange of traditions and cultural diplomacy has become a dynamo of change in international relationships. Fundamentally, this means that new powers are influencing and proposing diverse agendas that respond to their real internal and local needs. Cultural heritage is one of the captive elements of this new international agenda, and a close look at the new geopolitical world map, and even the cartography of world heritage sites considered by UNESCO, helps to consolidate this perception, which does not prevent a critical reading of such a cartography.

On the other hand, the pluralization of heritage narratives, which function as ambassadors for new world demands, has created increasingly challenging expressions of the Eurocentric legacy in the conduct of policies considered homogenizing the preservation of world heritage. In Africa, Latin America, Asia and the Middle East, criticism of the hegemony of these policies has given rise to distinct and emancipatory preservation actions, with the increasing presence of themes that address the "Africanities", "Latinities", "Asianities" and "orientalisms".

With the appropriation of cultural heritage for commercial and political purposes within the economies of all parts of the globe, heritage conservation now plays an important role in cultural diplomacy, elevating its status from a mere diplomatic strategy of good neighborly relations to an elaborate soft power tactic in different countries. Heritage has become an increasingly important theme of multilateral dialogues and, as such, has broadened its influence at the global level. These are just a few examples of the areas that bring together the preservation of cultural heritage and international relationships. Understanding the mechanisms for understanding this thematic expansion will favor the establishment of new heritage values, nationally and internationally.

1   Brazil on the Road to Illicit Trafficking in Cultural Goods

In Brazil, no specific legislation has yet been created to deal with the illegal trafficking of cultural goods and works of art, so the country still uses the basis of the 1970 UNESCO Convention Federal Constitution (1988) in addition to a fundamental set of laws such as Decree-Law No. 25 of November 30, 1937, which organizes the protection of the national historical and artistic heritage; Law No. 3924 of July 26, 1961, archeological and prehistoric monuments; Law no. 4845, of November 19, 1965, which forbids the export abroad of works of art and crafts produced in the country, until the end of the monarchic period; Bilateral agreement with Bolivia, of July 26, 1999, on the economic recovery of stolen cultural goods, illegally imported or exported; UNESCO Convention (1970): means to prohibit and prevent illegal import and transfer of the ownership of cultural goods, ratified on February 16, 1973, effective on May 16, 1973; International Institute for the Unification of Private Law (UNIDROIT), 1995, on stolen or illegally exported cultural objects.

It is worth noting that in 2008, the Brazilian Federal Police initiated a specific training and capacity building program for police officers in order to identify works of art and cultural assets in danger. The lack of specialization of the police in cases of trafficking in works of art and cultural goods makes the country an even easier gateway for these crimes. The trafficking of works of art and historical pieces, as well as of sacred art, puts our country's cultural heritage in constant danger. There are many pieces of Brazilian cultural heritage lost annually due to theft, robbery and illicit trafficking. For this reason, since 1997, together with the Federal Police, Interpol and International Council of Museums, Institute of National Historical and Artistic Heritage (IPHAN) has developed a campaign called "Fight Against Illicit Trafficking in Cultural Goods", whose objective is to return to the places of origin the stolen or illegally diverted works of art.

Stimulated by the greed of the receivers and the lack of security of the collections of some institutions and private collections, the thieves that operate in Brazil steal everything: statues, paintings, saints, books, documents, coins, photographs, maps, archeological pieces, fossils and any other type of object considered valuable for collectors. IPHAN's registration, which is legally responsible for the preservation of goods overturned by the federal government, registers more than a thousand pieces stolen throughout the country.

There is a scheme used by art theft specialists: once the theft is made, the pieces are distributed to rogue antique dealers who act as recipients. In the select commercial circle of these objects, the "collectors" are warned as soon as the new pieces reach the market. The next step is to sell them to private collections. From then on, the location of the work becomes practically impossible, because whoever buys, in general, knows that he/she is taking a stolen piece and does everything to hide the possession of the object. The stolen sacred pieces are resold mainly in the domestic market. The indigenous ceramics are in great demand abroad. Even Amazonian archeological urns are on the red list of International Council of Museums (ICOM) and are considered assets at risk of theft. Paintings by renowned Brazilian artists such as Cândido Portinari and Di Cavalcanti are particularly coveted and have been stolen in actions that the police know to be to order.

Despite the efforts made by the Department of Museums and Cultural Centers of IPHAN to instrumentalize the various institutions involved in the recovery of stolen works of art, cultural goods, historical and sacred art objects, IPHAN's extensive list proves that a significant portion of Brazil's historical and cultural heritage is missing. It should also be added that, in recent years, the illicit traffic of works of art and cultural goods has been more intense and, even when the stolen objects are recovered, investigations are closed with the arrest of the executors of the robberies, without reaching the receivers, antique dealers, museums, collectors and galleries, the real responsible ones for the crimes.

Starting in 2006, Brazil began to occupy a prominent place in the list of the ten countries with the highest volume of theft of cultural works in the world after the action perpetrated in the states of Rio de Janeiro, São Paulo, Minas Gerais and Bahia, in the collections of Itamaraty Palace, National Library and Chácara do Céu Museum in Rio de Janeiro. These crimes highlighted thefts of several objects and works of art, including paintings by Claude Monet, Pablo Picasso, Henri Matisse and Salvador Dalí. Since then, Brazil has suffered several lootings, such as those at Museu de Arte de São Paulo—MASP and at the Mário de Andrade Library, from where paintings by Cândido Portinari and Pablo Picasso were taken, as well as a set of old maps and other pieces of historical value.

All this without counting the baroque bootlegging of the old Brazilian colonial churches, which since the beginning of the twentieth century, have been systematically plundered from our territory and sold on the black market, which generates billions of dollars in illegal profit. The Database of Wanted Cultural Property, created by IPHAN, listed approximately 1,032 art objects stolen in Brazil in 1997 alone, without considering in this statistic the goods that were not inventoried and overturned by the Public Power. Two decades later, in 2016, the number reaches just over 57,000 missing objects.

As for the other countries, between 2000 and 2015, 453,000 art thefts were recorded in Italy alone, and in England the cultural losses represent a loss of approximately 600 to 750 million pounds per year. After Brazil entered this list of countries in evidence, and such thefts and lootings were reported internationally, the traffic of works of art and heritage gained evidence in the country. However, ten years after the Chácara do Céu Museum was stolen, the largest theft of works of art in Brazilian history, the works have not been identified or found, and the country remains on the FBI and Interpol lists as the country where one of the most important art thefts of the twenty-first century occurred (Askerud and Clément 2019; Costa and Rocha 2007, p. 264).

A curious and frightening fact is that, in 2015, Brazil was only behind the United States, France and Iraq in the world ranking of cultural theft. However, the partnership between IPHAN and the Federal Police/Interpol has already lasted 10 years and bears fruit. In February 2016, for example, IPHAN helped identify 40 books, some over a hundred years old, and 49 engravings seized in Argentina, which returned to the National Museum library collection. The law regulating IPHAN (Decree No. 25 of November 30, 1937) established that the owner of a fallen piece who fails to report the theft, robbery or loss to the Institute within 5 days is subject to a fine of ten times the value of the object. The attacks against the fallen goods are judged as crimes against the national patrimony. According to the Penal Code, anyone who destroys an artistic, historical or archeological asset is subject to a fine and can be arrested for a period of six months to three years. Since the 1980s, the institute has been developing the National Inventory of Furniture and Integrated Property (INBMI) program, which has technically identified about 90,000 items of fallen monuments. Most of these goods are of sacred art, and it is estimated that there are more than 400 thousand pieces to be inventoried.

Portuguese journalist Carlos Reis, in a 2006 report, discussed the potential damage that looting and theft cause to the countries involved. In an article entitled *Art Trafficking: The World Heritage Raiders*, Reis (2006) stated that the plundering and illicit trafficking of works of art cause irreparable damage to countries' cultural heritage. According to its notes, the European Union recognized that "illicit trafficking in cultural goods has reached such a scale that cultural heritage is permanently the target of significant and often irreparable damage, and there is a danger that the situation will get worse given the insufficient catalogues of cultural goods of public and private institutions" (Reis 2006). But what path should jurisprudence on trafficking take when the origin of the work trafficked or in the possession of museums that have practically institutionalized its possession is known? What should be done, besides the identification, public reprimand and forwarding of works and goods intercepted and/or in possession? The answer is complex, but necessarily involves a serious discussion about the return and/or repatriation of the works and/or cultural assets. This subject is still quite confrontational, but we will try to argue why the return shows itself as an acceptable way to repair the historical and cultural damages perpetrated in the last centuries to the countries that have seen their material heritage embellish museums, galleries and private collectors' halls around the world.

2 Return as Historical Reparation?

It is essential that the trafficking of cultural goods and works of art be addressed academically, as it is currently a gateway to several other issues that are extremely important for the safeguarding of cultural identity, since the "trafficking" element is consolidated as an active agent of the so-called *soft power*, vector of agreements and disagreements between countries. With this, the importance of repatriation/evolution to safeguard the history of various communities enters the agenda.

In the last decade, the publication of books, academic texts, government dossiers and newspaper articles on the subject has grown exponentially, especially in English language publications. From the immense collection produced in recent years, some critical texts stand out, such as those of Benhamou 2014; Brinkley 2013; Bokova 2016; Cosomano 2010; Costa and Rocha 2007; Cuno 2008, 2012; Edsel 2011, 2014; Frammolino 2011; Gross 2010; Howe 2014; La Follette 2013b; Nassif 2015; O'Connor 2012; Seif 2015; Veiga 2014; Waxman 2008, 2011; Wiziack 2015; Yates 2015, 2016, a collective on which this research is based.

The return of works of art to the countries from which the pieces originate and the repatriation of diverse cultural goods is a frequent subject among museologists, managers of cultural goods, curators, jurists of international law, enthusiasts of cultural decolonization from all over the world, especially among countries that have important works of their estate on display abroad.

The question of the return of unlawfully exported cultural goods is naturally more complex. According to Article 5, item 5, of the Convention of the International Institute for the Unification of Private Law (UNIDROIT), the request for return must—by analogy with the rules for the return of stolen objects—be filed within a period of three years from the moment the requesting State knows the location of the cultural object and the identity of its possessor, and in any case within a period of fifty years from the date of export or from the date on which the object should have been returned. The application shall be accepted if the competent authority recognizes that the conditions set out in Article 5, items 3 and 4, have been met, in particular on the cultural importance of the object for the requesting State.

In Latin America, this is the case with Mexico, Peru, Guatemala, the Andean and Central American countries, which have indigenous gold and silver pieces now being returned to their original museums by the United States and European countries. The examples do not remain only on the American continent. Egypt, Syria, Iraq, among several other nations, have been increasingly concerned about the repatriation of their archeological treasures, their works of art and their cultural assets. A few years ago, several of these museums in the U.S.A. and Europe began to return objects to their countries of origin, each case with its own history.

Although much attention is given to the act of repatriation itself, the return of the work does not always cause a commotion or retain greater significance in the countries and/or cultures that produced it, sometimes falling even into the limbo of oblivion and even becoming inaccessible to the public. Most Western museums now recognize the strong ethical sense of returning objects, especially in cases where these works have left their countries of origin under questionable circumstances. The case of the statue of the goddess Morgantina, bought by P. Getty Museum in 1988 for 18 million dollars, and returned to Italy in 2011 after the Italian prosecutor found out that it had been looted, illegally exported and sold by merchants who most likely falsified its origin, is just one of the thousands of contemporary examples.

Since the 2000s, the legitimacy of the permanence of goods taken under colonial domination and/or of important identity, memorial and ritualistic value of groups of non-European countries in European institutions has been increasingly questioned, and therefore there has been an increase in requests for repatriation and in restitution and reparation initiatives and in public opinion movements around this sensitive subject (Acerbi 2019, p. 45). To name but a few, the Colombian request to Spain for the return of 122 pieces of the Quimbaya treasury, which began in 2006; the Peruvian government's request to Yale University for the return of 46 pieces and 332 objects originating from Machu Picchu, which was diplomatically formalized in 2005 and led to the signing of a memorandum of understanding and cooperation in 2011; the Greek demand to the British Museum for the restitution of the Parthenon marbles, whose first demand took place in 1984 with the UNESCO Intergovernmental Committee, but which gained momentum after the crisis that hit the country in 2009. Also by way of illustration, the returns of Jacques Chirac and then Nicolas Sarkozy of Korean manuscripts, between the years 1990 and 2000; the performance of Rijksmuseum, in the Netherlands, which removes the colonial terminology from the titles and descriptions of the museum's works of art potentially offensive to other peoples, and also discusses the restitution of several pieces of its collection with Sri Lanka and Indonesia, among other countries of origin; the open letter sent in December 2017 by dozens of organizations and hundreds of personalities to the German government calling for the restitution of the colonial legacy of Africa present in Germany.

Some factors that we can list as promoters of this movement are the advances in post-colonial studies and conceptions, empowerment and greater visibility of marginalized and infra-national communities, greater international geopolitical insertion of southern countries, and a continuous process of erosion of paternalistic discourses of material impossibilities of transference from north to south in view of the more and more tangible possibilities of cooperative exchange to create or improve them.

The academic productions on Brazil's involvement in repatriation and restitution focus on those undertaken against the theft of cultural goods (Christofolletti 2017) and in the literature on the subject, terms such as restitution, return, repatriation and recovery are used sometimes as synonyms, perfectly interchangeable with each other, sometimes with different denotations—these, sometimes very specific, sometimes very comprehensive. It is relevant, therefore, that we read each writing or each treatise advised in view of the exact tone of each term used—or, on the contrary, its amplitude, its imprecision—also taking into account its gestation context, and the fact that each word chosen is a territory of dispute, fruit of delicate negotiation among agents with distinct intentions and positions. In our opinion, there are subtle differences between them, resulting from the semantics of words and their gradual construction in texts and practices.

According to Acerbi (2019, p.45), "recovery" and "repatriation", in the first place, place emphasis on the part that demands and, in a possible agreement, takes back or receives something back—be it given in a digital bank, any article of private or public property, or in the case of our study, the cultural assets. If we ask "who" to the verbs recover and repatriate, we have as the subject the requesting country. In this way, they convey to us a sense of movement, initiative and activity of this part more intensely than the other words; they value the plaintiff. The first one carries the idea of the recovery of the lost or the one who suffered some damage, of the continuation of a process after a break. Thus, for us, "recovery" means the action of recovering what belonged to a certain individual or society. The second of these terms shares this character, but in a less pronounced way. It defines the place or community that takes back as homeland, land of origin. The origin, however, does not necessarily link to legitimate possession, nor does it have to mean a national state. Therefore, we understand that "repatriation" is the action of bringing back to the place of origin a certain object, but not only, including alternative processes of "virtual repatriation" or circulation of the objects claimed or information about them that reverberates in the sense of reconnection between a group and its land, identity, religion, history, whether the process unfolds between different countries or different communities in the same country (Prott 2009, p. 23).

"Restitution", on the other hand, frames the issue from the side of the dispute to whoever is asked, the one who owns or holds the requested good and who makes the return to whoever demands it. "To return" is, therefore, to return something to its original state or to the one to whom it previously belonged. It is a delicate term, with specific legal contours in several national legislations, as in Brazil (In the Brazilian Civil Code, it is present 107 times, constituting a specific obligation defined between articles 233 and 246), and implies the non-acceptance of the possession put in check. It is not for nothing that, when the Statute of the Intergovernmental Committee for the Promotion of the Return of Cultural Goods to their Countries of Origin or its Restitution in the case of Illegal Appropriation was formulated in the 1970s, France and Germany wished to ban the term restitution, which was only used with the addition of its next condition—evidencing the intention to remove from the term itself its the character of evaluation of possession as undue, especially in a context still so sensitive to the emergence of independent states to resume their inheritance (Prott 2009, p. 21).

We consider restitution in an inclusive manner, not only of objective return of the objects requested, but also of action taken to repair the damage caused by the prolonging absence of an asset from those to whom it culturally belongs (Barkan 2009, p. 7). This is the term used in the 1970 UNESCO Convention, the 1995 UNIDROIT Convention, and the 2004 ICOM Code of Ethics, and it is common in the vocabulary of countries in their demands and in the expression of those who defend them (Acerbi 2019, p. 56).

"Return", finally, is the most generic and neutral term, which refers only to the movement that is undertaken when bringing or returning a certain object to a place where it at a past moment has been. Because of its scope, it is polyvalent, and can be mobilized in negotiations to varnish itself with diplomatic delicacy, in comments to maintain a more impartial posture before the problem or, as it will be in this text, simply to avoid repetition. The United Nations General Assembly resolutions on the subject always combine it with "restitution", side by side through an "or", as if to take a step backwards and let the member state decide which reading is more suited to the specific case in question. It is worth commenting that these terms and their different meanings privilege a vision that we need to bring to light here, but we want to deconstruct it, since it is based on categories of law that have often imprisoned the debate around the legality of possession or guardianship of property, its conditions of acquisition and maintenance and return. If we move (or add) (a) the axis of analysis to (include) cultural or ethical terrains, prioritizing concepts such as legitimacy and belonging, we relativize the imposition of binary or at least taxative rigidity of the terms listed above—as well as possession, property, inalienability, sovereignty—and open space to think about alternatives with other nuances, discussing shared heritage, rights of use, multiple appropriations, reflecting on and about more inclusive, participatory, plural (Kõnig) ways (de L'estoile 2018, p. 61). If we think that the core of the movement of requesting and returning cultural goods is the range of cultural, historical, religious, artistic meanings that they carry most prominently for those who claim them, we can understand that their effective presence, circulation and appropriation among the claimants, as well as reparation for the prolonged and unjust absence—before a legal seal of ownership—are the central objectives of return. This discussion is based on the Conclusão de Curso de Vitória dos Santos Acerbi: **O direito à arte e à história: repatriaação e restituição de bens culturais em disputa no Brasil,** defended in the course of History of UFJF, in 2019, under my guidance.

For some, repatriation, particularly of Western antiquities, refers to the persistence of a particular country in a globalized world. It is a kind of "stubbornness of objects", explains James B. Cuno, president and CEO of the J. Paul Getty Trust. In *"Who Owns Antiquity?"*, Cuno questioned whether certain museums have the infrastructure to safeguard the returned treasures—or keep them accessible, even away from the movement of major cities and capitals. Another significant example of the return of trafficked or purchased works (this time as a source of money laundering) is the repatriation of the works of former Brazilian banker Edemar Cid Ferreira, carried out by the U.S.A. courts and already mentioned in this text. On the website of Sotheby's auction house, there is no mention of the fact that the works belonged to the former banker and came from a judicial process, which in itself signals the little transparency of this type of trade carried out by auctioneers of international prestige. The canvas *Hannibal*, by New York artist Jean-Michel Basquiat, is another treasure that Brazil has brought back. The work, valued at 10 million dollars, belonged to banker Edemar Cid Ferreira, former controller of the Banco de Santos. Besides the piece, the statue *Togatus Romano* was brought back to Brazil, valued at 900 thousand US dollars. These were not the only pieces of Edemar seized by the Brazilian authorities, authorized by the Justice. In São Paulo, other canvases are exhibited in museums in the city.

Sharon Waxman, a reporter who has worked for two of the most important American newspapers, the *New York Times* and *the Washington Post*, has written a book whose title spells out parts of the nebulous equation of art trafficking in the world. In *Loot: the battle over the stolen treasures of the ancient world, the* author pointed out investigations carried out in Egypt, Turkey, Greece and Italy, showing the routes and schemes of trafficking in ancient works of art. Waxman stated that, among other procedures, the French used dynamites to release the zodiac from the Temple of Hathor in Dendera, which currently rests prominently in the Louvre. Campaigns are currently underway to return to Egypt important pieces of its cultural collection, among which the return of the famous Rosetta Stone, which is now in the British Museum, the bust of Nefertiti, now in the Egyptian Museum in Berlin, and the Zodiac of Dendera, allocated in the Louvre.

On the other hand, the brutality with which the English seized part of the marbles that made up the Acropolis of Athens is yet another example of hot forged imperialism. For this reason, the defenders of the return of the objects base their premises on ethical precepts and argue that colonial thinking must be definitively overcome. The reporter claims that several museums in Europe and the United States have committed irregularities by incorporating new objects into their collections, especially by accepting donations. Significant examples are the prestigious Metropolitan Museum in New York and the J. Paul Getty Museum in Los Angeles.

It is paradigmatic that cases of misappropriation of cultural goods, works of art and historical pieces by museums have reached the police pages of newspapers and sensationalist news several times in the last decade. In an unprecedented move, Italy has promoted an international trial of the illicit trade in museum antiques in the United States, including the Metropolitan and the Getty. A striking example was one of *Metier*'s most important curators, the American Marion True, considered one of the most powerful, respected and requested art historians in the world, accused of belonging to a network that negotiated ancient art.

Investigators have accused Ms. True of making acquisitions with unscrupulous dealers. And colleagues seemed satisfied with her disappearance, as if one of the world's most important art historians deserved to be the only American curator to be brought to court. Ms. Marion True, curator of antiques at the J. Paul Getty Museum, was formally charged by the Italian Supreme Court with belonging to a network that traded stolen art. The curator has become the epicenter of a more common history than you might think. The world of art trafficking was opened in an interview the curator gave to the *Washington Post* (Edgers 2015).

Ms. True admitted to having recommended to Getty the purchase of works that she knew had been plundered. But she admitted it with one caveat: "If she knew where the work had been taken from, she would press for it to be returned", she says. On the contrary, many of her colleagues did little, if anything, to research the origin of the works. And none of them were charged. The lawsuit against Ms. True was catalyzed by searches of dealers and a massive leak of internal Getty documents, obtained by two reporters from the *Los Angeles Times*, and offers a rare glimpse into the too close relationships between museums, dealers and collectors.

The fact is that Ms. True, as a curator, should not have been held responsible for the museum's acquisitions. These purchases were made by the managers and Getty's direction, and in this plot, which looks more like a Hollywood movie script, Ms. True insists that she did not conspire with an illegal trafficking network, as the Italian prosecutors claimed, but she claimed to have acquired art for the Getty, which she knew had been stolen. Why wouldn't she? She is everywhere. "Art is in the market", she said, describing the Getty's acquisition policy. "We do not know where it comes from. And until we know where it comes from, it is better to be in a museum collection. And when we know where it comes from, we will deliver it." She was not dealing in stolen goods. She "rescued" art by sending it to big museums. The speech can and should be analyzed, and it shows how complicated is the world of acquisition, repatriation and illegal trafficking of works of art and heritage. According to museologist Cícero Almeida, servant of the Brazilian Institute of Museums (IBRAM), repatriation is an issue that has to be analyzed on a case-by-case basis, but he admits it is favorable. "It is necessary to take into account that colonized countries already have conditions to take care of their pieces, and repatriation generates the positive side of stimulating society to recognize their production", says the expert in an interview with *Agência Brasil*. (Interview of Cícero Almeida to reporter Isabela Vieira. Return of works of art to the countries of origin is the subject of an international meeting on museum protection. Agência Brasil, July 11, 2012. Available at <http://memoria.ebc.com.br/agenciabrasil/not icia/2012-07-11/devolucao-de-obras-de-arte-aos-paises-de-origem-e-tema-de-enc ontro-internacional-sobre-protecao-de-mu>. Accessed on 13 Jan. 2020.

3   In Brazil, Two Decades of Loss and Damage

Since 2006, Brazil has been on the list of the ten countries with the largest thefts of cultural works in the world. Therefore, in these 13 years the country has entered the list of the top countries in the illicit traffic of cultural goods. Database of Wanted Cultural Goods created by Institute of National Historical and Artistic Heritage,

IPHAN, listed in 1997 approximately 1,032 art objects stolen in Brazil. In 2019 it counted around 1700. In Brazil, the most recent and remarkable crimes involving the artistic and cultural heritage were perpetrated in the State of Rio de Janeiro, in the collections of Itamaraty Palace, National Library and the Chácara do Céu Museum, which occurred, respectively, in July 2003, July 2005 and February 24, 2006, and which remains one of the ten insolvable cases in the FBI's list.

Innovative actions were carried out by IPHAN, represented by the creation of a register of missing cultural assets, allowing any interested party to access it for consultation. The register also makes it possible to report criminals or provide information on missing pieces and eventually found, through the Internet.

The most relevant international instrument on the subject is the "UNESCO Convention on the Measures to be Adopted to Prohibit and Prevent the Illicit Import, Export and Transfer of Property of Cultural Property", dated November 14, 1970, and the "UNIDROIT Convention on Stolen or Illegally Exported Cultural Property", dated June 24, 1995, which should also be mentioned. In the case of bibliographic heritage, areas of confluence have helped to map more consistently the route and modus operandi with which traffickers of rare works operate in Brazil, especially in the subcategories: experiences of theft of rare books and Special Collections (maps, manuscripts, journals, iconographies and musical collections) and cases of mutilations of rare books and Special Collections.

In 2017, the list of missing works, of which literary works stood out, included at least 934 pieces listed by Institute of National Historical and Artistic Heritage (IPHAN), most of which were stolen from churches, mainly in the states of Rio de Janeiro and Minas Gerais. Reinforcing the actions of the Brazilian government in the sense of combating money laundering, the financing of terrorism and the illegal trade of works of art and antiques, as well as discouraging the use of these goods to conceal or hide the illicit origin of certain financial assets, has been a tonic sponsored by the consortium of Brazilian action agencies IPHAN/COAF/Ministry of Foreign Affairs/Ministry of Finance, among others, due to the difficulty of measuring their economic value.

IPHAN acts in the prevention of money laundering and financing of terrorism as an accessory regulatory and supervisory institution, since it defines the warning signs, applies sanctions in case of omission and supervises the registration by traders and auctioneers of the sector. However, this does not make IPHAN the regulatory body of the entire art market, nor does the Institute manifest itself about the economic value of goods in commerce—which is a function of the market—nor does it investigate activities considered suspicious, which is a responsibility of the bodies and entities of criminal prosecution.

One of the great recent achievements was the standardization of art and heritage dealers. According to the norm, dealers and auctioneers of works of art and antiques, in addition to registering with IPHAN's National Registry of Dealers in Works of Art and Antiques (CNART), must establish internal control methods aimed at preventing money laundering and terrorist financing. They are also obliged to keep their own register with the data of operations in amounts exceeding R$10 thousand and of the respective clients involved. The rule also requires them to communicate to COAF, through the Financial Activities Control System (SISCOAF), operations made in cash in excess of R$10,000, as well as operations that are considered suspicious by them. A novelty brought by the ordinance is the need for an annual declaration of non-occurrence to IPHAN, mandatory for all traders that do not declare any occurrence to COAF during the year. Another fundamental project is **Dialogues on illicit trafficking of cultural goods: interoperability of systems** that aims to provide greater interaction and connection between the various databases in the country and abroad. The amount of old objects that can feed the licit trade is limited and its quantity decreases every day. Therefore, to supply the growing demand, more and more illicit means of acquisition are being used, since there is no way to increase, from day to night, the number of authentic objects in circulation in the market.

Just as an example, another modality that has suffered substantially from the damage of trafficking is the Brazilian archeological heritage. The Chapada do Araripe Paleontological Heritage, located in an area of approximately 10 thousand km^2, which includes the states of Ceará, Pernambuco and Piauí has one of the most important paleontological collections on the planet and has stood out in the list of the most relevant examples. It is common knowledge that, for years, an international network of smuggling has been hindering the scientific exploration of this collection by Brazilian institutions. Millenary fish fossils are sold on the black market for up to fifteen U.S. cents each. Gangs recruit simple people from the region, known as "fishers", to remove the fossils, which will be sold to crossers for amounts of five to ten reais. According to information from the Federal Police, the international trafficking of fossils in Chapada do Araripe makes Brazil lose about R$ 7,000,000.00 per year. Fossils that leave the Cariri region for about US$10.00 can be sold abroad for US$ 1,000.00, depending on their size. It is estimated that only 40% of the fossils discovered in Chapada do Araripe still remain in Brazil. The Brazilian Society of Paleontology estimates that there are more than 70,000 fossils of the Chapada do Araripe in foreign collections, and only 3,000 in Brazilian collections.

4 Suggestions for a Less Stolen Future

In view of the density of information provided in this text, and the possible direct impacts on the maintenance of the material culture and identity of national culture, some measures are suggested so that in the future the damage caused by the gears of illicit trafficking in cultural goods can be minimized. Although it has not been the task of this text to cover the entire issue, it has sought to place the subject in a historical perspective in order to illustrate the foundations of what is currently known about the illicit trafficking of cultural goods and works of art, as well as the increasingly reported returns or repatriations of works and goods practiced by the countries of jurisdiction to the countries of origin. We can affirm that Brazil has a dense tradition on the subject, accentuated in recent decades by the country's entry into the not very select group of states that head the list of the most vulnerable to trafficking in cultural goods. More effective legislation and new inter-institutional mechanisms are possible ways for Brazil to resolve the chaos of illicit trafficking that multiplies each year. Aiming to collaborate with the discussion, a series of suggestions are listed below that, in our opinion, can help in the more effective conduct of the preservation of the Brazilian cultural heritage, as well as in the curbing of the illicit trafficking of cultural goods in the country.

The first point to be highlighted relates to the need for increasingly comprehensive disclosure of the current "National Inventory of Mobile and Integrated Goods", which, although not yet completed, currently accounts for over 90,000 pieces in seven states. The greater dissemination of this instrument can help in public policies to prevent and coerce trafficking. In addition to the dissemination and systematic feeding of this inventory, other actions are indicated that would help in the purpose described above.

(a) The need for a thorough and detailed mapping of the lack of security of cultural and religious institutions in the country, with 24 h monitoring;
(b) Creation of a unique catalog with a more comprehensive and accessible description and photos of the movable assets;
(c) The encouragement of museum collection overturns;
(d) The creation of a National Council, established by presidential decree and headed by Ministry of Justice, where the government actors involved, as well as representatives of the private initiative and the third sector, would participate in this entity;
(e) The creation of a Brazilian "red list"—a catalog managed by International Council of Museums (ICOM), which would identify cultural goods in danger and in vulnerable areas, in order to prevent illicit sales and exports. Incredible as it may seem, the country does not yet have this instrument.
(f) The implementation of a clear and articulated national policy on illicit trafficking of cultural goods. Although UNESCO has had a convention on the subject since 1970, which Brazil signed in 1973, there were no major actions to combat the problem, according to representatives of the international organization.
(g) The strengthening of national, Mercosur and Unasur committees on the subject;
(h) The implementation and systematization of what specialists on the illicit trafficking of cultural goods call the strengthening of the "five Cs"—Communication, Knowledge (exchange of knowledge about what it is and realization of a Brazilian list), Training (from customs to the collector, from the museum to the collector), Confidence in the organs and the system;
(i) The deepening of "interoperability of systems"—fundamental in the fight against trafficking, accompanied by the exchange of successful experiences in the field of information technology aimed at preventing and combating illicit trafficking in cultural goods;
(j) The arrest of the illegal excavations;
(k) The training of experts on trafficking in countries that are in conflict;
(l) Multiply exhibitions with stolen works to show them around the world, which would help make them "unsaleable";

(m) Draw up a "black list" of "havens of concealment" of heritage;
(n) And as Jean-Luc Martinez, current director of the Louvre Museum, suggested, the creation of "Refuge Museums", which protect goods from countries at war at the request of a sovereign state and "Museums of expropriated works", while works of art cannot return to their countries, would act as a means of preservation of this legacy today, almost lost.

All these actions would help to inhibit asset smuggling, illegal sale and misappropriation around the world. But the best way to curb trafficking and protect cultural heritage is still to do the hardest, but at the same time the simplest, way: prevent people from buying and selling illicit material. Although this premise may seem utopian, it is at the fundamental core of the mechanism that governs the wheels of trafficking.

From mapping to discretion, from recovery to repatriation, each piece of this puzzle needs to be enlightened, otherwise, we will still not know how to answer the questions already asked in previous discussions: on the one hand, who is interested in the dilapidation of our assets and, on the other hand, what is the moral cost of weakening the gears of trafficking? Reflecting on such questions may be the first step toward the deceleration of the pulleys of this harmful gear.

References

Acerbi VS (2019) Course Conclusion Work: The right to art and history: repatriation and restitution of cultural assets in dispute in Brazil, defended in the course of History of UFJF, in 2019, under the guidance of Rodrigo Christofoletti

Alford KD (1994) The spoils of world war II: the American military's role in the stealing Europe's treasures. New York, N.Y: Carol Pub

Askerud P, Clément E (2019) The prevention of illicit traffic in cultural property. A UNESCO handbook for the implementation of the 1970 Convention. UNESCO, Paris. http://www.lacult.unesco.org/docc/Manual_de_la_UNESCO.pdf. Accessed 9 Feb 2019

Barkan E (2009) Making amends: a new international morality? In: Prott LV (org.). Witnesses to History: a compendium of documents and writings on the return of cultural objects. UNESCO, Paris

Benhamou F (2014) Neoliberalism and French heritage policy in the context of globalization. Maney Publishing, Heritage and Society, vol 1, no 7, pp 47–56. https://doi.org/10.1179/2159032x14z.00000000018

Bokova I (2016) Combating illicit trafficking in cultural goods. Official Memorandum, UNESCO, 16 Feb 2016

Brinkley H (2013) MFAA: The history of the monuments, fine arts and archives program (also known as monuments men). BookCaps Study Guides, March 28, 2013

Christofolletti (Org) R (2017) Cultural Assets and International Relations: the heritage as a mirror of soft power. Santos, Leopoldianum

Christofolletti R (2017) The illicit trafficking of cultural goods and repatriation as historical reparation, in: Christofolletti, R. (Org), 2017. Cultural goods and international relations. Heritage as a mirror of soft power. Santos, Leopoldianum

Cosomano E (2010) How does the black art market work? *Super Interesting*, 282th edn. http://super.abril.com.br/cultura/como-funciona-o-mercado-negro-da-arte. Accessed 20 Aug 2016

Costa TP, Rocha JS (2007) The incidence of reception and illicit traffic of works of art in Brazil. Revista da Faculdade de Direito, São Paulo, vol 4, no 4, pp 263–282. Doi:http://dx.doi.org/10. 15603/2176-1094/rcd.v4n4p263-282

Cuno JB (2008) Who owns antiquity?: museums and the battle over our ancient heritage. Princeton University Press, Princeton, N.J, Woodstock, Oxfordshire [England]

Cuno JB (ed) (2012) Whose culture?: the promise of museums and the debate over antiquities. Princeton University Press, Princeton

Edgers G (2015) One of the world's most respected curators vanished from the art world. now she wants to tell her story. The Washington Post, Washington, August 22, 2015. https://www.washingtonpost.com/entertainment/museums/the-curator-who-vanished/ 2015/08/19/d32390f8-459e-11e5-846d-02792f854297_story.html. Accessed 6 Sept 2015

Edsel RM (2011) Hunters of masterpieces: saving the Western art from Nazi plunder. Rocco, São Paulo

Edsel RM (2014) Saving Italy: the race to rescue a nation's treasures from the Nazis. Rocco, São Paulo

Feliciano H (1997) The lost museum: The Nazi Conspiracy to steal the world's greatest works of art, reprint edn

Frammolino R (2011) Chasing aphrodite: the hunt for looted antiquities at the World's Richest Museum Hardcover

Gross M (2010) Rogues' Gallery: the secret story of the lust, lies, greed, and betrayals that made the Metropolitan Museum of Art. Broadway Books, New York

Horta ML (1996) Illicit trafficking of cultural goods: Brazil's situation. In: Illicit traffic of cultural property in Latin America. Icom, Paris

Howe JRTC (2014) Salt mines and castles: the discovery and restitution of looted European art. CreateSpace Independent Publishing Platform, North Charleston, SC

Johnston PF (1993) Treasure salvage, archeological ethics and maritime museums. The International Journal of Nautical Archeology

Kõnig V, de L'estoile B et al (2018) Les collections muséales d'art non-occidental: constitution et restitution aujourd'hui. Perspective, v.1, 2018. Availível em: http://journals.openedition.org/per spective/9059. Accessed 15 Feb 2019

La Follette L (ed) (2013a) Negotiating culture: heritage, ownership, and intellectual property. University of Massachusetts Press, Amherst and Boston

La Follette L (2013b) The Trial of Marion true and changing policies for classical antiquities in American Museums. Negotiating culture: heritage, ownership, and intellectual Property, Chapter 2

Merryman JH (ed) (2008) Imperialism, art and restitution, 1st edn. Cambridge

Miles MM (2008) Art as Plunder: The ancient origins of debate about cultural property, 1st edn. Cambridge

Nassif L (2015) Art market is used worldwide for money laundering. The All Brazil Newspaper. http://jornalggn.com.br/noticia/mercado-de-arte-e-usado-no-mundo-inteiro-para-lavagem-de-dinheiro. Accessed 20 Aug 2015

Nicholas LH (1996) Europe plundered: The fate of European artistic treasures in the Third Reich and the Second World War. Companhia das Letras, São Paulo

Nye JRJS (1990) Bound to lead: the changing nature of American power. Basic Books, New York

O'Connor AM (2012) The lady in gold: the extraordinary tale of Gustav Klimt's masterpiece, Portrait of Adele Bloch-Bauer. Knopf, New York

Prott LV (2009) The history of return of cultural objects. In: Witnesses to History: a compendium of documents and writings on the return of cultural objects. UNESCO, Paris

Reis C (2006) Art trafficking—The robbers of the world heritage. Overseas Missionary Vision

Seif A (2015) Illicit traffic in cultural property in Lebanon. In: Desmarais F (ed) Countering illicit traffic in cultural goods: the global challenge of protecting the world's heritage. ICOM, Paris, pp 65–82

Todeschini C (2007) The Medici Conspiracy: The Illicit Journey of Looted Antiquities-From Italy's Tomb Raiders to the World's Greatest Museums. Paperback

UNESCO (2011) The fight against the illicit trafficking of cultural objects: the 1970 convention: past and future. Paris: UNESCO, 15 and 16 March 2011. http://unesdoc.unesco.org/images/0019/001916/191606E.pdf. Accessed 23 April 2016

Veiga JMF (2014) International/national trafficking in works of art (theft and forgery): an approach to organized crime. Createspace Independent Publishing Platform, North Charleston, SC

Waxman S (2008) Loot: the battle over the stolen treasures of the ancient world. Times Books, New York

Waxman S (2011) Withdrawal. El arte de robar arte, Turner, Madrid

Wiziack J (2015) Creditors at the bank want paintings by ex-banker Edemar Cid Ferreira. http://www1.folha.uol.com.br/mercado/2013/11/1375838-credores-do-banco-santos-querem-quadros-de-ex-banqueiro-edemar-cid-ferreira.shtml. Accessed 20 Aug 2015

Yates D (2015) Anonymous Swiss Collector. 2015. http://www.anonymousswisscollector.com/bio. Accessed 20 Sept 2015

Yates D (2015) Illicit cultural property from Latin America: looting, trafficking, and sale. In: Desmarais F (ed) Countering illicit traffic in cultural goods: the global challenge of protecting the world's heritage. Paris, ICOM, p 3

Jesuit-Guarani Missions: UNESCO World Heritage Site in Brazil

Tobias Vilhena de Moraes and Pedro Paulo Abreu Funari

Abstract This chapter aims at presenting some aspects of the concept of the archaeological preservation of the remains of material culture and the role of the archaeologists during site management planning at the archaeological Jesuit-Guarani Missions, a UNESCO World Heritage site in Brazil. Located in the southern part of the South American continent (in the state of Rio Grande do Sul), the Missions were excavated for the first time and conserved from the beginning of the 1980s. Since then, many conservation projects have been carried out at the site in an effort to obtain a more accurate method to site management planning. In this chapter, we present the recent results of site management planning at the Jesuit-Guarani Missions World Heritage site and its implications to heritage legislation in Brazil.

Keywords Historical archaeology · Archaeological preservation · UNESCO cultural heritage · Ruins of Sao Nicolau · Jesuit-Guarani missions

1 Heritage

Heritage is a very intricate subject. Worldwide, it deals with the past as if it is in the present, and as people want it to be in the future. Past and future are thus at the heart of heritage and the people may be key to use it for freedom, social inclusion, diversity respect, and coexistence. Heritage however has a history not so much related to the people, as it has been forged in terms of nation state and imperialism, and control of populations. People refers to a polity, to power relations, while populations are the way a government tries and controls individuals taken as part of subjects to be managed, as numbers in statistics. Heritage as a national concern meant that UNESCO was and still is a collection of nation states.

T. V. de Moraes (✉)
Lasar Segall Museum/Brazilian Institute of Museums/IBRAM, São Paulo, Brazil
e-mail: tovilhena@yahoo.com.br

P. P. A. Funari
University of Campinas, Campinas, Brazil
e-mail: ppfunari@uol.com.br

© The Author(s), under exclusive license to Springer Nature Switzerland AG 2021
F. Lopes da Cunha and J. Rabassa (eds.), *Festivals and Heritage in Latin America*,
The Latin American Studies Book Series,
https://doi.org/10.1007/978-3-030-67985-9_14

History though is the result of not only social movements, such as for women's, civil, youngsters', indigenous, or minorities' rights, but also anticolonial, antiwar, for freedom of sexual behaviour, among others, so much so that there is an increasing challenge to intolerance and exclusion worldwide, particularly since 1945 and during the next decades. Brazil though, during the Cold War (1947–1989), resulted in military coups, dictatorships, repression, under national security ideology and population control in Latin America, including violence and killing. Brazil experienced harsh times. Heritage suffered as a result, at least as there was little concern for tolerance or coexistence, as it was clear by the motto: "Brazil, love it or leave it". Even so, social moves never disappeared and heritage benefited too. In the interstices of the regime, there has been a growing place for the people, considered in their diversity: natives, illiterates, enslaved, outcasts, infamous, and common people.

The Jesuit Missions heritage represents in a nutshell all the contradictions just mentioned. The Missions put together Jesuits, natives in a complex relationship, as well as Paulista conquerors (Tupi-speaking from Portuguese America or Brazil), Spanish people, enslaved natives, Africans and mixed people, among others. Interpreted as a communist Edenic society by some, as the ultimate prison by others, a much more complex and contradictory mix in fact, they were, since the late eighteenth century a mass of debris, until taken as heritage much later, particularly from the twentieth century. In this paper, we aim at discussing some aspects of this move from populations to the people, in their fertile variety.

2   The Beginning of Archaeological Research in the Missions

Located in the southern region of South America, the Missions were the scene of a fierce territorial dispute between Spain and Portugal during the seventeeth and eighteenth centuries. As a result of the dispute, the Missions represent one of the disputed historical sites between Europeans and the indigenous peoples on the Latin American soil (Gutiérrez 1982). A process of cultural transformation altered the way of life of these populations forever under the influence of Europeans, however, without causing the complete disappearance of the local traditional indigenous culture (Barcelos 2000; Curtis 1993; Custódio 1987, 2002; Furlong 1937, 1962; Gutiérrez 1987, 1992; Kern 1998).

Today, the material remains of this site have evidence of cultural blending that occurred there as both the Europeans and indigenous people were involved in sculpture, ceramic manufacture, and architecture with skill and technique. However, after the Treaty of Madrid (1752), which was signed between the Spanish and the Portuguese following the Guaraní War (1752–1756), the Jesuits were expelled as a result of bad civil administration. This eventually culminated in the destruction and abandonment of the Missions (Pesavento 1994; Shulze-Hofer and Marchiori 2008; Weimer 1993). When the Missions were finally abandoned between the end of the nineteenth century and the beginning of the first quarter of the twentieth century, they were sporadically visited by travellers who were searching for a glorious artistic and religious past. Before checking the impact of the more recent projects on the missions, the first attempt to implement a site management plan for the Missions is discussed.

During the early 1990s, the first archaeological research in the region was carried out with the assistance of the Church and the Tower of São Miguel and this was sponsored by the Federal government. However, the research was also supported by architects but there were no archaeologists and this led to the loss of archaeological data (IPHAN 1998; Mayerhofer 1947).

The first reports of São Miguel das Missões excavations were released in 1937 when the Federal government started to intervene in the management of the ruins following the founding of National Historic and Artistic Patrimony Service (SPHAN) in 1938. Father Luis Gonzaga Jaeger who worked for the Anchietano Research Institute carried out further archaeological research of the Missions at the end of the 1950s. Jaeger excavated the urban areas of the reductions of São Borja, São Luiz Gonzaga, and São Nicolau. In his publication, *À cata de Tesouros Jesuíticos* (Hunting for Jesuit Treasures), Jaeger made it clear that there was no specific method that was used in the field (La Salvia 1982, 1983a, 1983b). During this period, he tried to understand the extent of the influence of the reduction by excavating the interior of the Church of São Miguel das Missões. From the results of these first excavations, it became clear that the objective of what was to be researched revolved around almost exclusively on the interest of the professional engaged in the excavations. The excavations, therefore, did not produce any academic or scientific conclusions. At that time, the study of archaeology in the country was still in its infancy in the region. The study of archaeology was missing a systematic and a well-defined theoretical and methodological approach for all the research in the country (Fig. 1).

3   The Missionary Archaeological Heritage in the Late 1970s

Since the 1970s, some architects focused on the research of São Miguel, São João Batista, São Nicolau, and São Lourenço Mártir. José Saia, the representative of SPHAN for the southern region of the country, organized some excavations of the Missions. Saia's objective was to determine the extent of the archaeological remains so as to establish the real dimension of the missionary space. According to Arthur Barcelos (2000), several methodological mistakes were made in this initial period of research. More specifically, Barcellos believed that the location of structures and trench openings had damaged the archaeological substrate. Facts caused by the lack of expertise of professionals involved in activities in which the knowledge and protection of the archaeological substrate was going to take precedence, considering the fragility of the ruins.

Fig. 1 São Miguel das Missões. Photo by IPHAN-RS

4   A New Approach to Missionary Heritage

In the late 1970s and early 1980s, there was a significant change in the technical and administrative way of managing the Brazilian archaeological heritage sites (Funari 2003). From this time onwards, in addition to identifying, registering, preserving, and restoring the monuments, the Federal government wanted the management of the sites to benefit the people. One of the main objectives of this new concept was to appropriate cultural assets in the name of the "nation" and to devolve their management to the local communities, which were considered to be the authentic owners (Gonçalves 2002; Prous 1992). This approach was in line with the country's historical re-democratization process, which took firm steps against the military dictatorship that was in force since 1964 (Días-Andreu 2007; Fausto 2013; Fico 2004; Funari 1994a, 1994b, 1995, 1999; Gonçalves et al. 2013; Montes 1998; Orser 1992).

A site management master plan called "*Diretrizes para o desenvolvimento físico de São Miguel das Missões*" (*Guidelines for the Physical Development of São Miguel das Missões*) was developed in 1980 by the municipality of Santo Ângelo (then sub-district of São Miguel), Secretary of the Interior, SPHAN, and the Regional Development and Public Works (Custódio 2002, 2009). The purpose of the project was to establish a more effective way of managing the ruins, which in turn was to lead to the management of the surrounding areas. Following this, the extent of the cultural landscape was established as well as its geography and topography. The technicians that were involved in the project wanted to encourage the growth of the small town of São Miguel, which is the sub-district of the municipality of Santo Ângelo. In addition, they wanted to provide tourism infrastructure and define how cultural heritage would be managed in the future. The plan was later used as an instrument to create a "Missions Park" that would be part of the regional and local tourist attraction whereas at the same time used to manage the site.

In order to execute this plan, 5 zones of activities were defined. The most complex of them was the historical preservation, which encompassed the other Jesuit missionary ruins as well as their surrounding areas. The aim was to develop an impact-damping zone, which would protect historic structures against the future growth of the present city. In order to properly manage the architectural assets, recommendations were made by the project managers (Custódio 1987, 2002, 2009; Stello 2005).

However, when the *Guidelines* are analysed today, it is realized that the archaeological component was still unknown in its totality and complexity. As a result, technical precautions were never suggested to preserve the "archaeological zones" nor were archaeological studies proposed prior to the consolidation of the ruins, which would make it possible to manage the cultural heritage of the site (Kern 1994; de Menezes 1984; Reis 2005).

5 The Excavations in São Nicolau and the Historical Missionary Archaeology Project

Excavations began at the archaeological site of São Nicolau around the same time that research progress of the site was made. The work, coordinated by Fernando La Salvia, resulted in an agreement signed between SPHAN and the Undersecretary of Culture, Sports and Tourism. The main objective of this project was to carry out archaeological excavations in a section of the old Jesuit-Guarani reduction.

A large area of the mission space was excavated (approximately $4500\,\mathrm{m}^2$). Located in the centre of the present São Nicolau, a small town with little more than 5000 inhabitants, the ruins were buried underground and were concealed by vegetation. During the excavation, various structures were exposed: a church, college, hospital, wine cellar, a sewage system located at the back of the church, and the town hall. The excavation was extended to the periphery of the old urban area where the garden of the Mission (*La Florida*), part of the industrial areas (*silo* and *eiras*), and water channels were exposed. The need to care for the archaeological features that were now exposed generated great concern for the members of the excavation team. What to do with the revealed structures and how to protect them from anthropogenic factors, animals, and nature were questions that needed to be addressed. Years later, La Salvia himself warned that the excavation should have had continuity with the stabilization of the remnants and a process of conservation of the floors and evidence, which did not occur at the time. He did not know why this happened, but pointed out in a later document that this caused very great damage to the excavated area and practically its loss (La Salvia 1983a).

The problems that originated from this excavation promptly led to a change of focus on the future of archaeology projects and heritage management. For the first time, the need to preserve what was exposed becomes important. Heritage technicians and archaeologists could no longer distance themselves from the need to preserve archaeological remains. Above all, the study technique on material culture assets was to be modified in a way which did not compromise the physical and cultural integrity of the site. Therefore, every restoration would not only focus on one object but on the entire local landscape. The immediate consequence of the problems faced in São Nicolau was the elaboration of a project aimed at safeguarding the archaeological missionary cultural heritage. This initiative was sponsored by the Secretariat of Culture, Sport and Tourism of Rio Grande do Sul and the National Pro-Memory Foundation, together with Fernando La Salvia.

As a consequence of this action, it was necessary that there would be real ways of safeguarding archaeological, historical, and architectural material so as to use the evidence as tourist information (La Salvia 1983a, b). The São Nicolau archaeological excavations were a milestone in Brazilian archaeology both for the results ascertained and for the challenges faced. From then on, the meeting of different professionals trained in the field of material culture would be considered in the decision-making processes on missionary heritage.

Following the departure of Prof. La Salvia, a new project called Historic Missionary Archaeology (1985) was structured through a technical cooperation agreement between UFRGS, PUC-RS, and SPHAN. The project was coordinated by the technical team that included Arno Alvarez Kern and Pedro Augusto Mentz Ribeiro who were teaching undergraduate students. The institutional coordination of the project was done by an architect called Julio Curtis. In addition to publications which describe excavated and collected artefacts, interpretive studies began to be carried out as well. Until that time, the focus of research on the Missions was restricted to just a survey of indigenous occupation and colonization (Carle 1998; de Curtis 1993; La Salvia 1983a, b; Brochado et al. 1969; Brochado and La Salvia 1989). In 1988, Arno Kern sought a general explanation of the human occupation in the region with the new project. The intention was to help produce a true historical synthesis of the missions. As a field method, and based on the ethnic diversity in the region, traditional methods of understanding prehistoric and historic archaeology were simultaneously applied (Kern 1998. Large-scale excavations, much larger than that of La Salvia, were carried out on large grid squares according to the stratigraphic methodology proposed by Mortimer Wheeler. Large area excavations, however, followed Andre Leroy-Gourhan's methodology by careful decoupage. All actions in the field were defined by grids and the layout of existing architectural structures. The problems encountered throughout the excavation defined the methods that were used.

6   The Missions World Heritage Site Management Plan

After integrating a scientific approach to archaeological research projects, a new discussion on preserving the Missions began to be discussed. Contacts with institutions inside and outside the country began to be seen as a necessary element for scientific development. During the 1990s, agreements and exchange activities began to be signed between the *Brazil-US Workshop* (1993) and The University of Arizona, National Park Service (NPS), and IPHAN from Santa Catarina and Rio Grande do Sul. Workshops for technicians from both countries served to delve into the best ways that could be used to manage the Missions. The workshops resulted in the idea of establishing an archaeology laboratory and the construction of an institution that specializes in the study of the Missions.

During the 1990s, other projects were, however, carried out around the cultural landscape of the Missions. Some of the projects include the School Site International/Missions (SEI, Sítio Escola Internacional/Missões) of 1992 and the Integrated Valuation Programme (PIV, Programa Integrado de Valorização) that were, respectively, carried out in 1994 and 1998. The first project was based on the UNESCO Convention, associated with the protection of humanity's cultural heritage, mainly focusing on the need for training and the creation of a regional centre for scientific research. SEI allowed the training of a large number of professionals and carried out excavations on several archaeological sites of the Rio de la Plata basin (Kern 1994).

The focus ceased to be on one archaeological site as it began to cover the broad area that was occupied by the missionaries. Contrastingly, PIV involved archaeologists, IPHAN workers and university students to work on various archaeological sites.

The computerization of scientific research was widely used during this time in Brazil. In the Missions, a specific project called *"Computerization of the Archaeological Site of São Miguel Arcanjo"* was established with the support of IBM and the Friends of the Missions Association. A meeting on the Archaeology and Informatics which involved the participation of guests from MERCOSUL (the South American Common Market composed of Brazil, Argentina, Paraguay, and Uruguay) was also arranged and this led to the establishment of the Archaeology and Informatics bulletin, database, multimedia programs. and a website. This was coordinated by José Otávio Catafesto de Souza, Francisco Noeli, and Luiz Felipe Escosteguy.

At the turn of the millennium, archaeological studies were further refined. A better understanding of geological and environmental space becomes important. Between 2000 and 2004, a project coordinated by Prof. Carlos Henrique Nowaztki of Center of Studies and Research in Archaeological Geology (NEPGEA, Núcleo de Estudos e Pesquisa em Geologia Arqueológica) of Unisinos, Rio Grande do Sul, sought to understand the geology of the Missions. The initial objective of this research was to determine the origin of the rock extraction site(s) used to build the Church of São Miguel. In addition to the acquired knowledge, this information would be used for future restorations and if applicable, could be performed with the same kind of rock that was originally found there (Nowaztki 2004, 2007). In the end, besides the location of the old quarries, it was possible to identify old missionary roads.

At the beginning of the new millennium, the IPHAN team, which had been active on the Missions, was not able to consolidate the sites of São Lourenço and São João. In order to solve this problem, restoration work began to rely on the participation of archaeologists hired for short periods of time and coordinated by IPHAN. One of the studies was to protect and enhance archaeological sites of São Lourenço Mártir and São João Batista, which was carried out between 2003 and 2005.

At São João Batista, the consolidation of the ruins was promoted by the IPHAN architects and they were also supported by archaeologists José Otávio Catafesto and Vera Thaddeu. As a result of this project, hundreds of archaeological artefacts were recovered and a large amount of information about that settlement was accumulated. Field surveys included teaching researchers who would work in the Missions. The training programme was thus established to conserve and manage the Guarani Jesuit Missions. The training programme was promoted by UNESCO, World Monuments Fund (WMF), and IPHAN, and was implemented between 2003 and 2006. This programme was based on the need to promote international cooperation in order to integrate conservation efforts and experiences. During this period, several workshops were held to promote practical conservation activities of archaeological sites in Paraguay, Brazil, and Argentina. Situations and experiences of the various countries that have ruins were evaluated during the meetings. Afterward, joint guidelines were established for the continuation of integrated actions. The basic conservation manual for Guaraní Jesuit Missions, distributed in 2009, included all the studies on topics that were discussed.

As part of these continuous conservation actions and following the management policies of the archaeological heritage of the Missions, several projects were implemented in the first quarter of this century. One of the projects which was carried out between 2008 and 2010 sought to accurately identify and delimit the Missionary Fountain (Fonte Missioneira) area. The project was coordinated by IPHAN and the archaeologist Vera Thaddeu and, later by a company called Zanetinni Arqueologia. The Missionary Fountain had been accidentally found after a backhoe drilled the land to expand the area for cultivation. The stone structure, carved with care by craftsmen, contained three angel faces, through which the water flowed into a hole open at the mouth. The fountain was damaged and there was also an attempt to depredate the site. In order to control the damage, architects restored the fountain while José Otávio Catafesto carried out a study of the area. This study led to the discovery of an old channel that carried water from a natural source close to the Fountain.

A new project, carried out in the 2000s, had as the main objective the preservation of the archaeological remains in the Missionary Fountain Park area so that the local residents and tourists were able to better enjoy it. After two work campaigns, new structures were discovered. They comprised a complex system of local water supply with pipelines and water tanks. The research made it possible to precisely delimit the areas of interest and of probable new archaeological findings.

It is important to highlight that these last projects were all inspired by the accomplishments attained by the missionary historical archaeology project, which pointed out the need to focus on the research area. Even with the researcher's attention turned to a study focused on the initial projects, the investigations were characterized by a concern to compose a general panorama of the occupation. As an example, one can look at the studies on Guarani indigenous ceramics as an ethnic identifier or the work on the use of metal in the Guarani-Missionary world (Kern 1998).

In 2002, IPHAN signed an international cooperation agreement with the Andalusian Institute of Historical Heritage (IAPH-Spain, in Portuguese, Instituto Andaluz do Patrimônio Histórico). The main objective of this agreement was to carry out an extensive study of the regional cultural landscape in the territory of Sete Povos das Missões, which involved various scientific fields such as archaeology, history, architecture, and anthropology. The project integrates several areas of culture such as intangible heritage and missionary statuary, and it aimed to recover and enhance the missionary archaeological heritage complex. This agreement established as one of its main items the creation of cooperative relations of a scientific, technological, formative, and cultural nature in relation to documentation, conservation, formation, and dissemination of historical heritage (Informe 2007). The project was titled Cultural Landscape Guide for the Management and Development of the Jesuit Missions Territory in Brazil. The project, however, focused more on the application of geophysical and archaeological surveys to understand the extent of the Missions (Fig. 2).

Fig. 2 São Miguel das Missões. Photo by Luiz Bolcato Custódio/IPHAN-RS

7 The Missionary Management: From Past to Present

As noted in previous paragraphs, over the past 30 years several archaeology projects have been carried out on mission sites. These projects focused mainly on preservation activities associated with architecture and restoration. The presence of archaeologists was often voluntary and temporary and the objective was to recover built structures.

The various field studies brought together a large number of professionals, including professors and students from several universities in the state of Rio Grande do Sul who intervened in the numerous missionary archaeological sites with the intention of understanding more deeply the history of the missions. The material found was collected, and the data was taken to laboratories and reserves where it was treated and placed on shelves and never to be released to the public again. Only researchers interested in the subject could access it.

The seasonal interest in the area, the lack of continuous projects, and the vision of archaeology as a mere auxiliary science of restoration made possible, at most, two large research projects of greater importance that involved professionals and the community simultaneously. The organ responsible lacked a permanent technical body of mainly archaeologists in the missionary sites.

The same factors that hindered the formation of a preservation and scientific programme coupled with the cyclical movement of people with other customs and culture, which in general do not show great interest in satisfying the curiosity of the local people, created a certain awkwardness, if not an almost complete indifference and repudiation to the archaeologist in the local communities. This problem is not very different from that faced in other parts of the world with eloquent archaeological heritage such as Mexico and Italy where there is still a certain difficulty in reconciling knowledge and dissemination in the research agenda, with the participation of the outside community.

Archaeology is becoming increasingly concerned with the social and public aspects that encompass its work, trying to enhance the ways it interacts with society (Holtorf 2007). Archaeology assumes an interpretative role in ancient societies. One of the objectives of this science is to understand the material culture associated with human beings. At the same time, more and more, they note that communities have a sense of belonging to their own past and want to participate in decision-making processes. In this way, these communities have their own idea of asset appreciation (Holtorf 2005; Little 2007; Menezes 1984; Poulot 2009).

This change of perspective in archaeology began in the 1970s, as well as the idea that many histories may exist in one place. From the North American archaeology perspective, this new vision emerges with the Civil Rights Act (1964), when a hitherto essentially white people view of the landscape is seen as "multiracial", leading the responsible bodies to make more inclusive interpretations of the past (Shackel in Shackel 2004). Different individual and group interests began to be addressed in order to understand the meaning of material culture. Sociocultural anthropology emerges from this new branch of archaeology, from the discussion of dialogue and collaborative work. However, as it became clear from analysing several projects of public archaeology, it is necessary to deepen the knowledge of all participants about the past while the thecreation of bonds is developed.

In the missions, as a way to overcome the community's lack of knowledge of their own archaeological heritage, and to minimize distrust towards archaeologists, heritage educational actions involving the community were developed as early as in the Fonte Missioneira project. Didactic booklets and a map of the city which show archaeological sites were created and distributed in the schools of the region. A routine of visits to the excavation sites was established by local scholars and lastly, mini-courses were created to teach professionals who work with the archaeological heritage in the region.

Nonetheless, the absence of a long-term plan is one of the main factors that lead to the successive discontinuous actions of archaeology. The fluctuation of institutions and projects involved in the sites, over the last 30 years, is a reflection of this lack of continuity.

There were some attempts at continuity, such as the "Cultural Tourism Program" (Programa de Turismo Cultural) for the missionary region. The programme focused on areas of cultural and environmental tourism, conservation of structures, museology and museography, implementation of visitation itineraries, and interpretation, signage, and diffusion of the remnants of the former settlement. The participation of tourism in the development of the site promoted an opportunity to elongate the study, with visitors being able to contribute to the socioeconomic increment of the region (IPHAN 1998).

However, the archaeologist's role in that programme was practically restricted to the disclosure of archaeological remnants (such as the Indians house, church, and wine cellar) of the four sites. It was not clear what technical-scientific criteria would be adopted to "expose", or rather to value each asset. In addition, it appears that the archaeologist researcher would also integrate routine actions employed by IPHAN with their work.

8 Not One, but Several Projects of Management Planning

As defined in the introduction, the initial objective of this article is to analyse how the numerous archaeological projects were created and their relationships with the local communities.

For this reason, the development of the concepts of preservation, archaeology, and restoration was systematized; the archaeological excavations that occurred in the remnants were identified and organized; several educational actions were observed; and, lastly, an analysis was elaborated according to the precepts of archaeological preservation dictated by the patrimonial letters and by the work experience gained in the Missions.

According to this analysis, in the last 30 years, several researchers and managers have tried to permanently include archaeology as an everyday work tool for the management of archaeological heritage in their projects. These attempts have faced a number of issues over time, such as shortage of funding and logistical support, as well as a struggle in interpreting the work process between archaeologists and architects.

As discussed earlier, the insertion of archaeology into restoration projects is recent. Especially in Brazil, where, in the years 2000, in Porto Alegre, Recife, and Bahia, archaeologists protested and demanded the definitive presence of archaeologists in the field, as was already required by law. This triggered the creation of a Federal programme called the "Monumenta Program" (Programa Monumenta).

Until that time, the presence of the archaeologist field researchers only happened sporadically. The very lack of a preservation policy and a technical structure capable of managing that project was a strong hindrance to planning projects. IPHAN, for instance, had a small technical staff, with only five archaeologists trained to operate in all national territories.

The alternative found since the 1990s was to outsource several of IPHAN's activities such as project management, coordination, and execution. The Institute would perform routine activities such as project analysis and inspection, a tiny job in the face of such great responsibility.

This situation distanced fieldwork professionals from the field of preservation (archaeologists, architects, historians, etc.) and reinforced the temporary and fleeting nature of cultural heritage protection activities.

However, as it became clear throughout this article, the national criteria that guided IPHAN throughout its more than 70 years to the presence or absence of continued programmes during the restoration projects was not an obstacle to carrying out the activities under their responsibility on a regional scale. That is, institutions and professionals from the state of Rio Grande do Sul were able to articulate various actions for the valorization of that cultural heritage.

Characterized as a national agency that operates at the state level through its regional offices, IPHAN is usually shaped according to regional specificities. As Ton Ferreira and Suely Amâncio (2011) pointed out in an article about the Monumenta Program in the state of Sergipe, it was evident that the mere analysis of the legislative framework (laws, decrees, ordinances, etc.) would be insufficient and would not account for the local reality (2011). It would be necessary to understand how laws, techniques, and theories have been adapted in a regional context.

From this perspective, it was possible to perceive that the continuous formation of archaeologists and architects in the state of Rio Grande do Sul and the technical experience gained from the continuous excavations in the Jesuit-Guarani Missions helped to consolidate the practice of preservation and restoration in the national territory.

Although archaeology has only recently played a major role in Brazil, it was able to contribute—albeit for a short time—to reduce the distance between the cultural asset and part of the community, whether through educational activities or through knowledge produced before and after the excavations carried out over the years.

This is not to say that the main challenges have been overcome and nothing more remains to be done. As discussed at the beginning of this conclusion, frequent discontinuation of research activities is a major challenge. In addition, a lack of daily activities to reconcile the various social agents (politicians, indigenous people, farmers, etc.), the communication between the various agencies that work in the missionary sites (IBRAM, City Hall, State Government, etc.), and, lastly, a more adequate socialization of archaeological assets for those who are today excluded from visiting the site (people with special needs, indigenous people, etc.) proved to be a challenge.

9   Suggestions for the Future Management Planning

This article could not end without introducing some propositions for the future of management planning in the Jesuit Missions. All of them are offered to take into consideration the discussions presented in this article on the development of the concept of archaeological preservation in Brazil, and the numerous studies of the ruins of this important Latin American cultural legacy. These propositions may serve as a first step in overcoming the limits imposed so far on missionary cultural heritage management projects.

Over the last four decades, the projects that were carried out had the main objective of protecting, preserving, and disseminating the history of the Jesuit-Guarani

Missions. Considering the experience gained through managing such projects and, nowadays, the theoretical and critical support of information acquired from field research and reflection such as this article, some observations can be made to possibly strengthen the effectiveness of planning and management of those heritage sites.

- Implementation of a federal public contest to employ a technical archaeologist in the Missions;
- Implementation of a public contest to employ archaeologists in the municipalities of the mission region where there are sites listed;
- Elaboration of a sensorial visit tour to archaeological sites focused on promoting the well-being and development of the person with some sort of disability;
- Qualification and/or construction of an Archaeology Laboratory and Technical Reserve in the Missions.

Throughout this trajectory, heritage has been gradually incorporated into the scope of the missionary archaeological site management programmes. Several heritage public policies were continuously establishing a project on the existing cultural assets in the Brazilian missions. This position helped to consolidate the concept of integrated work, that is to say, with the participation of professionals from different areas, reminiscent of the initial proposal of the missionary historical archaeology project and even that of La Salvia in the early 1980s, which gathered colleagues of different professions, specialized in the preservation of cultural assets, in which the archaeological patrimony is inserted (Barretos (Org.) 2008; Bastos and Funari 2008; Funari et al. 2005; Little 2007; Murray and Evans (Orgs.) 2008).

Even Kern (1994, 1995, 1998 and other publications) exposed his thoughts several times on the necessity of combined efforts on the archaeological, architectural, and document recording processes. Similar positions that proposed, for instance, the joint analysis of history sources, anthropology, and archaeology could also be found in foreign researchers of the same period such as Deetz (1988). Kern's (1998) initial attempt was to improve flawless preservation archaeology practices and to integrate the necessary sources of research (Lima 1993, 2001, 2002).

At any given time, archaeologists as well as different professionals such as anthropologists, architects, geographers, forest engineers, museologists, and educators contributed to the discussion: an intentional movement to approximate different sciences that sought to gather information about the local cultural heritage complex.

The initial archaeological excavation work in the Missions was a milestone in Brazilian archaeology both for the results ascertained and for the challenges faced. Going forward, the meeting of different professionals trained in the field of material culture and the need to preserve the archaeological assets were going to be the cornerstone in the construction of decisions on the cultural missionary heritage (Basu 2008; BRASIL 1993, 2006; Bueno and Machado 2003; Gonçalves 2002; Gonçalves et al. 2013; Shanks and Tilley 1993; Poulot 2009; Symanski 2009). From now onwards, the ultimate goal would be to have comprehensive and complete information on each of the recovered objects, the actual size of the sites and their interrelationships, and finally, to allow the archaeological studies to support the projects of management of cultural assets (Días-Andreu 2007). A new stage for missionary research projects has begun (Funari et al. 2005; Funari and Pelegrini 2006; Funari and Silva 2007; Kern 2002).

The Mission offers a unique opportunity to deal with heritage, as the complex relationship of present, past, and future (Fig. 3). The present is always with us and it is also a gift, a present, as we are alive. Without a past, there would be no present. The future, as the past, is an abstraction in the present, and the future is nothing, except

Fig. 3 São Miguel das Missões. Photo by Eneida Serrano/IPHAN-RS

for the possibilities open by our own wishes and actions for pleasure. Heritage may serve to close doors, or rather to open them. It may foster compliance or freedom, depending on our choice, and the Missions may serve different and even opposite goals. We argue in this paper that it may serve a better, more equal, and peaceful coexistence. We will be happy if this chapter may contribute to that.

References

Barcelos AHF (2000) Espaço e arqueologia nas missões jesuíticas: o caso de São João Batista. Porto Alegre: Pontifícia Universidade Católica do Rio Grande do Sul (EDIPUCRS), vol 600

Barretos (Org.) EA (2008) Patrimônio cultural e educação: artigos e resultados. Universidade Federal de Goiás, Goiânia, p 2008

Bastos RL, Funari PPA (2008) Public Archaeology and management of the Brazilian archaeological-cultural heritage. In: Silverman H, Isbell WH (Orgs.) Handbook of South American Archaeology. Springer, New York, pp 1.127–1.133

Basu P (2008) Confronting the past? Negotiating a heritage of conflict in Sierra Leone. J Mater Cult Lond 13:233

BRASIL (1993) Ministério da Cultura 1993. Manual de gerenciamento do patrimônio arqueológico. Ministério da Cultura/IBPC, Rio de Janeiro, p 1993

BRASIL (2006) Ministério da Cultura. Iphan. Coletânea de leis sobre preservação do patrimônio, Rio de Janeiro

Brochado JP [et al] (1969) A cerâmica das Missões Orientais do Uruguai, um estudo de aculturação indígena através da mudança na cerâmica. Arqueologia da Área do Prata. In: Simpósio de Arqueologia da área do Prata e adjacências. Instituto Anchietano de Pesquisas, São Leopoldo

Brochado JP, La Salvia F (1989) Cerâmica Guarani. Posenato Arte e Cultura, Porto Alegre

Bueno LMR, Machado JS (2003) Paradigmas que persistem: as origens da arqueologia no Brasil. http://www.comciencia.br/reportagens/arqueologia/arq16.shtml. Accessed 12 April 2014

Carle CB (1998) O conhecimento dos metais nas Missões. RS-Brasil. In: Kern A Arqueologia histórica missioneira. EDIPUCRS, Porto Alegre

de Curtis JN (1993) O espaço urbano e a arquitetura produzidos pelos Sete Povos das Missões. In: Weimer G (ed) A arquitetura no Rio Grande do Sul. Mercado Aberto, Porto Alegre

Custódio LAB (1987) Missões, uma história de 300 anos. Porto Alegre: Iphan/ 12ª CR/Comissão Missões

Custódio LAB (2002) A redução de São Miguel Arcanjo: contribuição ao estudo da tipologia urbana missioneira. 199 f. Unpublished Master thesis, Universidade Federal do Rio Grande do Sul, Porto Alegre

Custódio LAB (2009) Missões jesuíticas: arquitetura e urbanismo. Cadernos de História, n. 21. http://www.memorial.rs.gov.br/projetos-cadernos.htm. Accessed 5 Jan 2009

Deetz J (1988) American historical archaeology: methods and results. Science 239:362–367

Días-Andreu M (2007) Internationalism in the invisible college: political ideologies and friendships and archaeology. J Soc Archeol 7(1):29–48

Fausto B (2013) A vida política. In: Schwarz, L. M. (dir.). Olhando para dentro 1930-1964. Objetiva, vol 4, São Paulo

Ferreira LM (2002) Vestígios de civilização: a arqueologia do Brasil imperial (1838/1877). Unpublished Master thesis, Universidade Estadual de Campinas, Campinas

Ferreira LM (2010) Território primitivo: A institucionalização da arqueologia no Brasil (1870-1917). EdiPUC-RS, Porto Alegre

Ferreira T, Amâncio-Martinelli S (2011) O Programa Monumenta e a problemática da aplicação da arqueologia na restauração dos monumentos históricos brasileiros. CLIO Arqueologia, UFPE 26(1):21–47

Fico C (2004) Além do golpe: versões e controvérsias sobre 1964 e a Ditadura Militar. Ed. Record, Rio de Janeiro

Funari PPA (1994a) Rescuing ordinary people's culture: museums, material culture and education in Brazil. In: Stone PG, Molyneaux BL (eds) The presented past—heritage, museums and education. Routledge, London, pp 120–135

Funari PPA (1994b) South American Historical Archaeology. In: Historical Archeology in Latin America. The University of South Carolina, Columbia

Funari PPA (1995) A hermenêutica das ciências humanas: a história e a teoria e práxis arqueológicas. Revista da SBPH, Curitiba, n. 10:3–9

Funari PPA (1999) Western influences in the archaeological thought in Brazil. In: World Archaeological Congress, vol 4. South Africa

Funari PPA (2003) Arqueologia. Contexto, São Paulo

Funari PPA, Orser Jr C, Schiavetto (Orgs.). SNO (2005) Identidades, discurso e poder: estudos da arqueologia contemporânea. Annablume; Fapesp, São Paulo

Funari PPA, Pelegrini SCA (2006) Patrimônio Histórico e Cultural. Jorge Zahar, São Paulo

Funari PPA, Silva GJ (2007) Gladyson José da. "Nota de Pesquisa sobre o Projeto de Pesquisa do Acervo de Arqueologia e Patrimônio de Paulo Duarte. História e História, pp 1–25

Furlong GC (1937) La arquitectura en las misiones guaraníticas. Estudios 57, Buenos Aires

Furlong GC (1962) Misiones y sus pueblos de Guaraní. Balmes, Buenos Aires

Gonçalves JRS (2002) A retórica da perda: os discursos do patrimônio cultural no Brasil. UFRJ/MinC-Iphan, Rio de Janeiro

Gonçalves JRS, Guimarães R, Bitar NP (eds) (2013) A alma das coisas. Patrimônios, materialidade e ressonância. Maud/FAPERJ, Rio de Janeiro

Gutiérrez R (1982) La misión jesuítica de San Miguel. Dana, Cedodal, no 14

Gutiérrez R (1987) As Missões Jesuíticas dos Guarani. Fundação Pró- Memória/Unesco, Rio de Janeiro

Gutiérrez R (1992) Arquitectura y Urbanismo en Ibero-América. Cátedra, Madrid

Holtorf C (2005) From Stonehenge to Las Vegas: Archaeology as popular culture. Altamira Press, Oxford

Holtorf C (2007) Archaeology is a brand. Archaeopresse, Oxford

IPHAN (1998) Programa de Turismo Cultural. Relatório de Grupo Interdisciplinar de Trabalho. IPHAN, Brasília

Kern AA (2002) O futuro do passado. Os arqueólogos do novo milênio. Trabalhos de Antropologia e Etnologia (Sociedade Portuguesa de Antropologia e Etnologia), Porto, Portugal, vol 42, no 1–2, pp 115–136

Kern (Org.), AA (1994) A arqueologia e o Sítio-Escola Internacional do curso de pós-graduação em História da PUCRS. Veritas 39(154):199–209

Kern AA (1995) A carta internacional da Arqueologia ICOMOS. SAB, Porto Alegre

Kern AA (1998) Arqueologia Histórica Missioneira. EDIPUCRS, Porto Alegre

La Salvia F (1982) Evidenciação, interpretação e ambientação dos remanescentes das antigas missões jesuíticas no Rio Grande do Sul. Secretaria de Cultura, Desporto e Turismo, Porto Alegre

La Salvia F (1983a) São Lourenço Mártir: algumas idéias para uma pesquisa arqueológica. Revista Ciências e Letras da Faculdade Porto-Alegrense de Educação Ciências e Letras, Porto Alegre, no 3, pp 67–75

La Salvia F (1983b) A Arqueologia nas Missões e uma perspectiva futura". In: Simposio Nacional de Estudos Missioneiros, 5, 1983, Santa Rosa. Anais. Faculdade de Filosofia, Ciências e Letras Dom Bosco, Centro de Estudos Missioneiros, Santa Rosa

Lima TA (1993) Arqueologia histórica no Brasil: balanço bibliográfico (1960–1991). Anais do Museu Paulista

Lima TA (2001) A proteção do patrimônio arqueológico no Brasil: omissões, conflitos, resistências". Revista de Arqueologia Americana, México 20:53–79

Lima TA (2002) Os marcos teóricos da Arqueologia Histórica: possibilidades e limites. Revista Estudos Ibero-Americanos, Porto Alegre, v. XXVIII, n. 2:7–23

Little BJ (2007) Historical Archaeological: why the past matters. Left Coast Press, California, Walnut Creek

Mayerhofer L (1947) Reconstituição do Povo de São Miguel das Missões. Tese de Concurso. UFRJ, Rio de Janeiro

de Menezes UB (1984) Identidade cultural e patrimônio arqueológico. Revista do Patrimônio Histórico e Artístico Nacional, n. 20:33–36

Montes AM (1998) Considerações sobre a restauração arquitetônica em arqueologia. Tradução: Iphan/Deprom. 1998. 83 f. Unpublished doctoral dissertation, Escola Nacional de Antropologia e História, Cidade do México

Murray T, Evans (Orgs.) C (2008) Histories of archaeology: a reader in the history of archaeology. Oxford University Press, Oxford

Nowatzki CH (2004) O sítio arqueológico de São Miguel das Missões: uma análise sob o ponto de vista da geologia. All Print, São Paulo

Nowatzki CH (2007) A Geologia Arqueológica na Unisinos. Cadernos IHU Ideias, São Leopoldo, ano 5, no 75

Orser CE Jr (1992) 1992. Oficina dos Livros, Introdução à arqueologia histórica. Belo Horizonte

Pesavento SJ (1994) História do Rio Grande do Sul. Mercado Aberto, Porto Alegre

Poulot D (2009) Uma história do patrimônio no Ocidente: séculos XVIII-XXI. Estação Liberdade, São Paulo

Prous A (1992) Arqueologia brasileira. UNB, Brasília

Reis JA (2005) Um palimpsesto sobre teoria na Arqueologia Brasileira. Arqueologia Sul-Americana, vol 1, no 1

Shackel PA (2004) Working with communities: heritage development and apllied archaeology. In: Shackel PA, Chambers EJ (eds) Places in mind: public archaeology as apllied anthropology. Routledge, New York

Shanks M, Tilley C (1993) Re-constructing archaeology. Routledge, Cambridge

Shulze-Hofer MC, Marchiori JNC (2008) O uso da madeira nas reduções jesuítico-guarani do Rio Grande do Sul. Iphan, Porto Alegre

Stello VF (2005) Sítio Arqueológico de São Miguel Arcanjo: avaliação conceitual das intervenções 1925–1927 e 1938-1940. Unpublished Master Thesis (Mestrado em Engenharia Civil)—Universidade Federal do Rio Grande do Sul, Porto Alegre

Symanski LCP (2009) Arqueologia Histórica no Brasil: uma revisão dos últimos vinte anos. In: Morales WF, Moi FP (eds) Cenários regionais de uma arqueologia plural. Annablume/Acervo

Weimer GA (1993) 1993. Mercado Aberto, Arquitetura no Rio Grande do Sul, Porto Alegre

"Invisible Heritage": New Technologies and the History of Antarctica's Sealers Groups

Andrés Zarankin and Fernanda Codevilla Soares

Abstract This paper discusses alternative forms of heritage construction and preservation related to the human occupation of Antarctica by subaltern groups and its invisibilities in the official discourses on the colonization of the continent. Our proposal associates public, digital and sensorial archaeology approaches, highlighting a more pluralistic and democratic narrative about the southernmost past. For this purpose, we used tools such as new technologies applied to archaeological research (3D laser scan, object scanner, 3D printers, drones and others). Besides, through itinerant sensory exhibits (which simulates the Antarctic environment within an inflatable dome), we seek to narrow communication channels between the archaeological and non-archaeological public. Also, we encouraged the construction of multivocal narratives on the human occupation of Antarctica.

Keywords Antarctic archaeology · Mediation · Technology · Sensory stimulation

1 Introduction

In 2009, the project "Landscapes in White: Antarctic Historical Archaeology" began, which aims to understand the processes and human strategies of colonization of Antarctica over time (Zarankin and Senatore 2007; Zarankin et al. 2011). This is a continuation of the Project of Historical Archaeology of Antarctica initiated in

A. Zarankin (✉)
Department of Anthropology and Archaeology, UFMG. Coordinator, Laboratory of Antarctic Studies in Human Sciences, Faculty of Philosophy and Human Sciences of the Federal University of Minas Gerais, Avenida Presidente Antônio Carlos, 6627, Sala 3070, Pampulha, Belo Horizonte, Minas Gerais CEP 31270-901, Brazil
e-mail: zarankin@yahoo.com

F. C. Soares
PNPD/CAPES Researcher, Laboratory of Antarctic Studies in Human Sciences, Faculty of Philosophy and Human Sciences, Federal University of Minas Gerais, Avenida Presidente Antônio Carlos, 6627, Sala 3070, Pampulha, Belo Horizonte, Minas Gerais CEP 31270-901, Brazil
e-mail: codevilla2005@hotmail.com

Argentina in 1995, coordinated by Dr Andres Zarankin and Dr Maria Ximena Senatore. The research emerged from a tri-national partnership between the Antarctic archaeology and anthropology teams from Chile, Argentina and Brazil, constituting a joint effort that, instead of establishing rivalries, allowed a collaboration that materialized in an advanced and unique study at the world level. Coordinated by the then recently created Laboratory of Antarctic Studies in Human Sciences (LEACH) of the Federal University of Minas Gerais—UFMG, the project was included in the research sponsored by PROANTAR (Brazilian Antarctic Program) and CNPq. Thus, Brazil incorporated, for the first time, studies in Human Sciences within its Antarctic program. In the specific case of our project, the objective was initially to study the first human strategies of Antarctica occupation, between the end of the eighteenth century and the beginning of the nineteenth century, centred on South Shetland Islands. Later, we expanded our research by incorporating a line of anthropological analysis aimed at thinking about Antarctica from a broader viewpoint, through macro processes, which have granted it different identities through time.

In 2010 we carried out the first fieldwork directed by LEACH, excavating archaeological sites in Byers Peninsula of Livingston Island, in the South Shetland Archipelago. The sites had already been identified and georeferenced by the Argentine team in the 1990s (Zarankin and Senatore 2007).

In the 24 years of research development, new actors have been inserted in the narratives produced, highlighting the daily life of the sealer, sea-wolf hunter and whaler, from the late eighteenth century and early nineteenth century, in the history of the region. In this specific work—combining public, digital and symmetrical perspectives—we propose new forms of human relations with Antarctica, seeking the involvement of the non-archaeological public with research in development. In a parallel and complementary way, we intent the construction of another Antarctic heritage, centred on subaltern groups, thus opposite to that of the official history. For this reason, besides analysing the history of groups without history (Wolf 1982)—sealers, sea-wolf hunters and whalers—we will also discuss new ways to insert the non-archaeological public as agents of these narratives. To this end, we will make use of digital technologies as a mediation form between them.

This work included the use of technological tools in archaeological research and, on the other hand, the realization of a touring sensorial exhibit about Antarctica. In the first proposal, we used the tools such as three-dimensional digital scans of sites, 3D prototypes of artefacts, videos, visual records of drones, the database and the LEACH website, gathering an unprecedented and rich material collection, already with open access or with access in development. The second proposal consisted in the assembly of an inflatable dome, which simulated an Antarctic archaeological site and brought the visitor closer to the sensation of being in the Antarctic continent. To acclimatize the interior of the dome, we used equipment to reproduce the sounds of animals and storms, lighting to recreate the luminosity and brightness, data show for videos about the animals and the landscapes of Antarctica, and air conditioning to simulate the coldness of the region. The activity was developed in partnership with the Pedagogical Center (CP) team at UFMG, being an auxiliary action to the Itinerante Ponto Museum at UFMG, resulting in thousands of visitors.

2   Brief Historical Background of the Sealers in Antarctica

We must take into account the global context of the late eighteenth century, in which the presence of people in this continent is related to the dynamics of capitalist expansion, to understand the dynamics of human incorporation in Antarctica. This incorporation is not done by nations seeking to demarcate their sovereignty, but by companies seeking economic gain, which were simultaneously exploring different parts of the world. In this approach, human presence in Antarctica lands was directed towards a determined logic and formed part of an economic strategy. This economic strategy can be compared to the process for the incorporation of other marginal areas into the system, for example, the islands of the Indian Ocean (Richards 1982), southern Patagonia and South Atlantic islands (Silva 1985; Senatore 1999), according to expanding policies motivated by economic issues.

These companies were exploiting occasional resources whose marketing offered high profits. The distance and difficulty of access offered possibilities of little competition and generated expectations of high economic performance. The effort to obtain "profit" based on an equation between cost and benefit drives the system. Thus, these ventures, carried out by companies, were increasingly extending to the limits of the known and exploited. The incorporation of Antarctica into the capitalist dynamic consisted of expanding the scope of action of these companies. For its part, the exploitation of animal resources also followed this logic. The discovery of other colonies of marine mammals and new waters for hunting cetaceans resulted in a greater abundance of derived products—oils and hides. Therefore, this caused a saturation of the market and, consequently, a fall in prices at the time. To maintain the income, the companies had to increase the volume of exploitation, which led to the indiscriminate hunting of marine mammals, drastically reducing the populations in the newly incorporated areas. This reduction and the high cost of access to southern latitudes caused the companies to retract their scope for action. In this way, the capitalist "logic" itself prevented the extinction of these species, since the balance between costs and benefits reversed. At the end of a few years (ca. 1820-1825), partly due to over-exploitation, this activity ceased to be profitable and, thus, the companies changed their sights to other territories.

Consequently, the human presence in Subantarctic Islands and Antarctica in the eighteenth and nineteenth centuries is episodic and responds to fluctuations in time that are related to changes in the profitability of this activity. From isolated historical data, it is possible to affirm that during the nineteenth century there were three major cycles of exploitation of these southern spaces related mainly to the exploitation of sea-wolf hides (O'Gorman 1963; Martinic 1987). These last incursions may have diversified the strategy, extending the range of resources exploited from elephant oil or sea-wolf and seal leather and various cetacean products.

The impact of these first occupations on the islands of the South Shetland Archipelago is on the map, which shows the sites of sealers and whalers (Fig. 1).

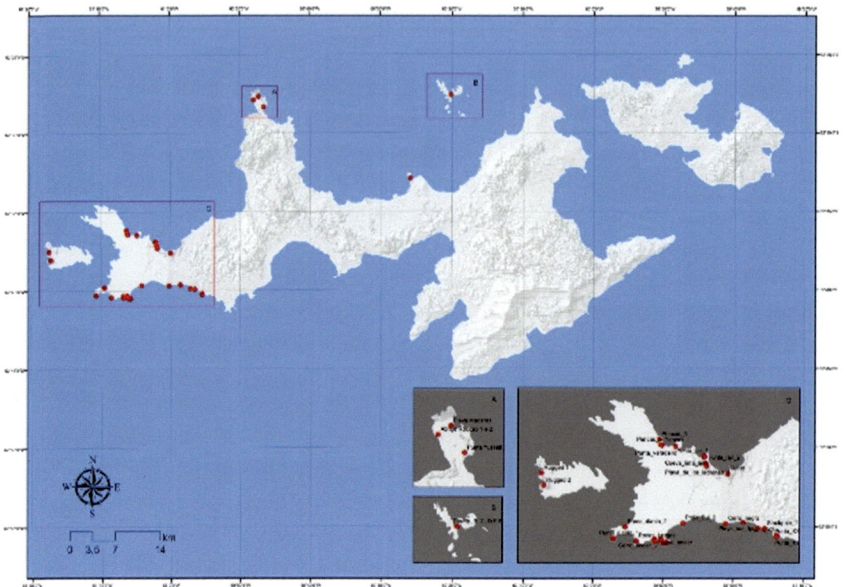

Fig. 1 Archaeological map of South Shetland. *Source* LEACH, 2014

3 For a History of People Without History in Antarctica

The official history has been concerned with telling the history of "heroes" or "important men", keeping ordinary people and subordinate groups invisible throughout this process, especially the sealer and whaler groups, protagonists of these first occupations (Senatore and Zarankin 2011). Archaeology, due to its disciplinary characteristics, allows the construction of an alternating history of marginalized groups from the material culture (in this case, remains associated with the daily life of these people in the South Shetland Islands).

Among the antecedents that led us to the archaeology of people without a history in Antarctica, we highlight the archaeological investigations in the 1980s developed by the Chilean team led by Rubén Stehberg (Stehberg and Nile 1983; Stehberg and Lucero 1985a, 1985b). These researchers excavated different stone shelters in the coastal fringe of Cape Shirreff (Livingston Island) and in the May 25th Island (Stehberg and Cabeza 1987; Lucero and Stehberg 1996). Interpretations have been published, pointing to isolated findings (projectile tips and a human skull identified as Amerindian) made on one of the islands (Torres 1992). All indicated that these findings needed new explanatory marks. In the 1990s, new lines of research in Antarctic archaeology were opened. Through cooperative agreements, Spanish archaeologists, under the direction of Martín-Bueno, joined the Chilean team and expanded the scope of the initial project (Martín-Bueno 1996a). Underwater archaeology is incorporated to locate the remains of a Spanish ship whose shipwreck is closely related to one of the hypotheses of the discovery of South Shetland—the ship San Telmo. This team explored different points of the coasts of the islands with different techniques and located possible wrecks (Martín-Bueno 1995a, 1996b). Also, in the 1990s, the Argentine team (of which Dr Andrés Zarankin was the coordinator, together with Dr María Ximena Senatore) was incorporated into the study of this problem, generating new information, which hinted at the magnitude of the incursion in certain places of South Shetland Islands. Antarctica was beginning to find other stories that remained unwritten.

Thus, the space of South Shetland Islands began to be systematically explored from archaeological research. They were part of the known areas: Livingston Island, Cape Shirreff (Stehberg and Nile 1983, Stehberg and Lucero 1985a, 1985b) and Byers Peninsula (Zarankin and Senatore 1999, 2000), May 25 Island (Stehberg and Cabeza 1987, Lucero and Stehberg 1996) and Rugosa Island (Pearson and Stehberg 2006).

In Byers Peninsula only, on Livingston Island, more than 30 archaeological sites have been recorded, most of which refer to the stone shelters corresponding to seasonal camps, linked to the first human exploitation of marine resources. From the excavations carried out at different sites, their functionality has been determined as productive spaces and shelters. The organization of the different camps presents a marked diversity, although linked to each other (Zarankin and Senatore 1999). The associated artefacts were dated in the late eighteenth century and early nineteenth century (Moreno 1999; Soares et al. 2016, 2017), consisting of remains of clothing, work tools and activities of daily life. Aspects of the daily practices of people who temporarily occupied these lands were studied (Senatore and Zarankin 1999). The use of local raw materials was identified both for shelter and the manufacture of various artefacts (Senatore and Zarankin 1997). The investigations made so far have worked as a starting point for the development of a new project, within which Brazil leads a collaboration with Chile, Argentina and, currently, also Australia and the United States, ensuring the integration of the information generated and the production of new data, which will allow advances in the knowledge about the sealers in Antarctica.

4 Antarctic Heritage

Until recently, only the remains associated with explorers and adventurers such as Scott, Amundsen, Shackleton, Mawson, Nordenskiöld and others were considered Antarctic historical heritage. This line of interpretation "preserves", mainly, the histories of exploration through the celebration of events, specific dates and protagonists, in certain places of space (shelters). In contrast, other histories linked to resource exploration were forgotten and silenced. It should be said that the histories of sealers and whalers have no specific protagonists, no exact dates and no facts of "historical relevance" to commemorate. However, they have left a large amount and diversity of material remains dispersed in the South Shetlands and the continent, which have a scarce place in the Antarctic cultural heritage conservation agenda. From the material culture, the history of these sealer and whaler groups can be told. Thus, the conservation of these traces is fundamental to preserve their history.

5 An Interactive Proposal for Building Knowledge of Antarctic Archaeology

We share the idea that COMMUNICATION (or mediation) is an integral, necessary and indispensable part of the archaeological discipline (Bezerra 2012; Sabloff 2008; Lemos 2014). More than informing the results of the research (by unidirectional means), our intention is that the non-archaeological public draws meaning from its contact with Antarctic Archaeology, building knowledge in an active and participative way.

Merriman (2004: 12) recalled that "the central issue of communication and interpretation [in archaeology] is the agency of the public and the degree to which experience is permitted to form and guide public engagement."

Thus, we are working with a focus that encourages the non-archaeological public to take ownership of Antarctica's history and create their interpretations of the traces and sites. We do not seek to correct "misconceptions" about the occupation of Antarctica or to suggest appropriate facts about the history of the continent. Our intention, as opposed to this, is that the public (archaeological and non-archaeological) have access to the different tools and information produced throughout the project. Therefore, they will evaluate, from various forms and sources, the history of Antarctica, allowing them to construct, in a diverse way, ideas that explain the human presence on the continent.

We understand that the actions developed have the potential to evaluate archaeological practice and how it has contributed to the construction of "patrimonies", highlighting mediation and communication as essential tasks of research and encouraging multivocality in the production of plural discourses on the polar past. In this sense, the Blank Landscapes project becomes a fertile substrate which inserts the public to transform and build knowledge about the continent. This construction occurs through a complex mediation process, in which the actors (human or not) participate actively and engage in the formulation of decentralized and multiple narratives. In this way, knowledge about the Antarctic continent can transcend the walls of the academy and the pages of books, being appropriated and built by the general public.

6 Digital Technologies and Sensory Exhibit

As part of this kind of knowledge, collaborative and symmetric, the LEACH team started the three-dimensional scans of Antarctic sites with the help of the Leica P20

3D Laser Scan, which produces digital, georeferenced and coloured versions of the sealers' shelters in the shape of point clouds (Fig. 2) (Soares and Mota 2017; Soares et al. 2018).

This data is processed by the Cyclone 9.0 software package, which helps to remove noise and produce accurate information about the scanned landscapes. The digital sites are transferred to a software, 3D Reshaper, capable of generating polygonal and textured models, which will be inserted in digital environments such as Google Earth or made available on the laboratory website.

It is intended that, from the three-dimensional models made available online, visitors can access Antarctica by touring the sites according to their intentions, creating routes themselves within the sealers' shelters and sharing, interactively, information about the traces and these human groups from hyperlinks, images or videos.

The laboratory's website contains a georeferenced database in which information on Antarctic archaeological sites, as well as the materials analysed, will be available on our website: www.leach.ufmg.br. It is available in Portuguese, Spanish and English.

The Database had as a starting point the elaboration of a system that would assist in the collection of information in the field, available for use through mobile devices, such as tablets and smartphones (Fig. 3). This system includes forms that help in the work of excavation, identification, conservation and registration of Antarctic sites. Besides being open for completion, the forms show information from other campaigns, making available data, in real time, that we would only have access to when in the laboratory. We hope that this tool will help in taking decisions in the fieldwork, making researchers more engaged with the excavation work and expanding the interpretative possibilities available *in loco*.

The Database is being produced in Open Source format and through languages made available by the Modi Research Group of Columbia University (USA), which is already used by The Earth Institute—Columbia University. The university provides an online database server, where it is possible to maintain a secure and reliable data storage to stash the information collected. This system will be integrated with Google Maps and Google Earth tools, and will also be available online by open access.

Fig. 2 Use of Leica P20 3D Laser Scan in fieldwork in Antarctica (*Source* LEACH Collection, photographer Michael Pearson, 10/01/2016), highlighting the point clouds generated by the three-dimensional digital scanning of the archaeological site Sealer 4 (*Source* LEACH Collection, 2016)

Fig. 3 Use of tablets and database in fieldwork (*Source* LEACH Collection, photographer Jimena Cruz, 12/01/2015)

So far, the database contains field and laboratory datasheets (Fig. 4). The field records are subdivided into five categories: for surveying, recording, excavation, conservation and analysis. The laboratory data sheets comprise individual sheets for each trace category, that is, we have specific datasheets for glass, ceramics (subdivided into crockery, kaolin and ceramics), metals, wood, lytic, cork, sediment, charcoal, fabric, leather and others.

Besides, during and after the analysis of the materials in the laboratory, we have also performed the scanning of material traces using MakerBot Digitalize and other types of three-dimensional modelling, using the Fusion 360 software produced by Autodesk, as well as photogrammetry. Once the models have been produced, they are textured, distributed or printed for the production of matrices used in the production of physical versions of the digital reconstructions of the traces (Fig. 5).

The three-dimensional models of the artefacts have helped actions of conservation of the vestiges, as well as archaeological analysis activities, especially those related to the measurements of the artefacts. Three-dimensional prototypes are also available online, on the LEACH website and Sketchfab, for public access.

The replicas of the traces are reproduced in physical versions using the 3D MakerBot Replicator 2 × printer. These materials are integrated into didactic processes of presentation of Antarctic archaeology and help restore the collection.

About the didactic activities, the project has carried out a sensorial exhibit about Antarctica, in which visitors are invited to immerse themselves in an inflatable capsule, which has, inside, a life-size replica of an Antarctic site (Punta Elefante

SÍTIOS

+ Cadastrar sítio   Q Buscar

Exibindo **1-20** de **32** itens.

Sigla	Nome do sítio	Pesquisador	Ano da campanha	Data de registro
Lima-Lima	Cueva Lima Lima	Fernanda Codevilla Soares	2010	18/09/2018
CS1	Cerro Sealer 1	Luisa Valadares Vieira	2010	30/10/2018
PX2	Punta X-2	Fernanda Codevilla Soares	2010	01/08/2018
PX1	Punta X-1	Luisa Valadares Vieira	2010	27/11/2018
PX3	Punta X-3	Luisa Valadares Vieira	2010	27/11/2018
CN	Cerro Negro	Luisa Valadares Vieira	2010	27/11/2018
S1	Stackpole 1	Luisa Valadares Vieira	2010	25/09/2018
S2	Stackpole 2	Luisa Valadares Vieira	2010	30/10/2018
PS1	Playa Sur 1	Luisa Valadares Vieira	2010	27/11/2018
PV1	Punta Vetor 1	Luisa Valadares Vieira	2010	27/11/2018
PV2	Punta Vetor 2	Fernanda Codevilla Soares	2010	27/11/2018
PV3	Punta Vetor 3	Fernanda Codevilla Soares	2010	01/08/2018
CS3	Cerro Sealer 3	Luisa Valadares Vieira	2010	18/09/2018
CS4	Cerro Sealer 4	Luisa Valadares Vieira	2010	01/08/2018
CS2	Cerro Sealer 2	Luisa Valadares Vieira	2010	27/11/2018

Fig. 4 Field and laboratory datasheets from the LEACH—UFMG georeferenced database (*Source* LEACH Collection, 2016)

II site) and artefacts such as bottles, pipes, vertebrae and whale ribs (among others) (Fig. 6).

Besides being able to visualize and handle these traces, visitors are invited to feel the cold and wind of Antarctica, listen to the sound of marine animals and participate in an activity embodied over the continent, since the inflatable dome is entirely adapted to stimulate these sensations.

The proposal was thought from the understanding that we are incorporated beings and our experience of the world is sensory, so the experience of sites, landscapes and our relationship with archaeological objects cannot be restricted to a visual or descriptive activity (Pellini 2012). According to Pellini (2016), the senses help in the creation of a world image, from them we can discuss how social groups structure or have structured reality in a certain socio-cultural context. This approach understands that the body is the foundation of human existence and, from it, reality is perceived and built (Tilley 1994). Thus, the senses concern, in the first place, our corporal engagement with the world (Howes 1991, 2003; Classen, 1993, 1997). Therefore, seeing, listening, touching and liking (among others) are not only ways to learn physical phenomena, but also ways to transmit cultural values (Howes and Classen 2009) and knowledge.

To apply this understanding to Antarctic public archaeology works, we believe it is fundamental to simulate an Antarctic environment, using an acclimatized dome, in which the archaeological and non-archaeological public can experience the sensation of being on the continent.

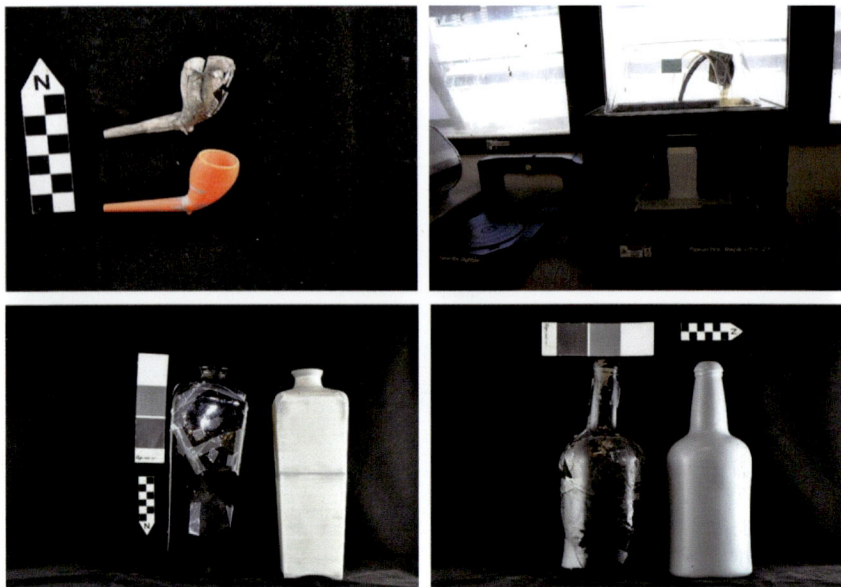

Fig. 5 3D Printer from LEACH—UFMG, and next "original" kaolin pipe and its replica (*Source* LEACH collection, photographer Matheus Mota, 05/06/2016); photo sequence (from left to right): 3D printing of troncopyramidal bottle (*Source* LEACH Collection, photographer Matheus Mota, 03/20/2016); "original" troncopyramidal glass bottle and its replica (*Source* LEACH Collection, photographer Matheus Mota, 03/25/2016); "original" cylindrical glass bottle and its replica (*Source* LEACH Collection, photographer Matheus Mota, 03/15/2016)

This activity, in a specific way, distances itself from the classic museum exhibits, merely exhibiting a visual that tries to transmit information, where the visitors assume the role of data receptors and not of knowledge builders. To opposing this classic approach, the sensorial dome stimulates a kinaesthetic experience about Antarctica, simulating what it would be like to be in the continent and feel, in an approximate basis, the cold, the sounds (sea, animals, wind noise and voices of researchers excavating), the brightness and observing the images (photos and videos) of animals and landscapes of the austral end.

In this sense, emphasizing the experience of the public, we understand that the built perceptions on the Antarctic past and present will be multiple, divergent and dissonant. Our proposal is, precisely, to stimulate these different understandings about Antarctica, starting from the body experience and digital archaeology as a means of interlocution and communication.

We are still preparing evaluation actions for all these activities, which will occur permanently at all stages. The purpose is to analyse whether the results obtained throughout the work are in line with the objectives initially proposed and to see progress and difficulties and also to redirect actions if necessary. Thus, the evaluation assumes a guiding, cooperative and interactive sense.

7 Final Considerations

The actions presented in this work seek to use new technologies as a way to develop the preservation of the material traces of subordinate groups in Antarctica and stimulate and improve the dialogue between us and the non-archaeological public for the construction of another heritage and history of the region. Examples such as the use of the sensorial dome show the potential of a space in which the public can become an agent in the construction of knowledge about the human occupation of Antarctica; in the same way, digital technologies can insert us in the chains of mediation to transform archaeological work in unexpected ways.

Fig. 6 Sensory dome of Antarctica; photo sequence—from uppermost and clockwise direction: external part of the sensory dome of Antarctica (*Source* LEACH Collection, photographer Fernanda Codevilla Soares, 09/19/2016); internal part of the sensorial dome of Antarctica with a replica of Punta Elefante II site (*Source* LEACH collection, photographer Fernanda Codevilla Soares, 09/19/2016); Internal part of the sensorial dome of Antarctica with a replica of the traces used in didactic activities (*Source* LEACH Collection, photographer Fernanda Codevilla Soares, 09/19/2016); visitation by elementary school students (*Source* LEACH Collection, photographer Fernanda Codevilla Soares, 09/19/2016)

Acknowledgements We thank the Itinerante Ponto Museum of UFMG for the partnership signed and actions developed, Matheus Motta for the help in the application of new technologies in the project and Prof. Fabiana Lopes da Cunha and Prof. Jorge Rabassa for the invitation to participate in this volume.

References

Bezerra M (2012) Signifying heritage in Amazon: a public archaeology project at Vila de Joanes, Marajó Island, Brazil. Chungara 44(3):363–373
Bonacchi C (2012) Introduction. In: Bonacchi C (eds) Archaeology and digital communication. Towards strategies of public engagement. Archetype Publications, London, pp 9–15
Bowles C, McCuistion A, Means B (2013) Virtual artifact curation of the historical past and the next engine desktop 3D scanner. In: Technical briefs in historical archaeology, vol 6, pp 1–12
Classen C (1993) Worlds of sense. Routledge, New York
Classen C (1997) Foundations for an anthropology of the senses. Int Soc Sci J 153:401–20
Dobres M, Robb J (2000) Agency in archaeology: paradigm or platitude? In: Dobres, MR (orgs) Agency in Archaeology. Routledge, London & New York, pp 1–18
Figueiroa R (2012) Por uma arqueologia das mídias: digitalizando em 3D o acervo cerâmico do museu de arqueologia de Xingó [For a midia archaeology: 3D scanning of the pottery collection from the Xingó Archaeological Museum]. In: 2° Colóquio de História e Arte História, arte e religiosidade nos caminhos da educação, 2012, Recife. URL: http://www.encontro2012.rj.anpuh.org/resources/anais/15/1338211118_ARQUIVO_ArtigoAnpuhRio2.pdf. Accessed 19 April 2016
Hodder I (2000) Agency and individuals in long-term processes. In: Dobres MR, Robb J (orgs) Agency in archaeology. Routledge, London & New York, pp 21–33
Howes D (1991) The varieties of sensory experience: a sourcebook in the anthropology of the senses. University of Toronto Press, Toronto
Howes D (2005) Sensescapes: embodiment, culture and environment. In: Classen C (ed) The Book of Touch. Berg, Oxford
Howes D, Classen C (2009) Doing sensory anthropology. www.sensorystudies.org/?page_id=355. Accessed 29 Nov 2016
Kellner A (2010) Mistério sob o gelo—Uma aventura na Antártica [Mistery on the ice: An adventure in Antarctica]. Rocco, São Paulo
Latour B (1994) Jamais fomos modernos. Ensaio de antropologia simétrica [We have never been modern]. Editora Nova fronteira. Rio de Janeiro
Latour (1994b) On Technical mediation—philosophy, sociology, genealogy. Common Knowledge 3(2):29–64
Latour (2012) Reagregando o social. Uma introdução à teoria do ator-rede [Reassembling the social. An introduction to Actor-Network-Theory]. EDUFBA e EDUSC, Salvador
Lemos C (2014) Se me der licença, eu entro; se não der, eu vou embora: Patrimônio e Identidade na comunidade quilombola Chacrinha dos Pretos (Belo Vale/MG). [If you excuse me, I'll come in, If you don't, I'll leave: Heritage and Identity at the quilombola community Chacrinha dos Pretos]. (Dissertação de Mestrado). Universidade Federal de Minas Gerais. Belo Horizonte. 146p
Maddison B (2014) Class and colonialism in Antarctic exploration, 1750–1920. Pickering and Chatto Publishers, United Kingdom
Machado M, Brito T (orgs) (2006) Antártica: ensino fundamental e ensino médio [Antarctica: elementary school and high school]. Ministério da Educação, Secretaria de Educação Básica, Brasília
Merriman N (2004) Public archaeology. Routledge, London

Pellini J (2010) Mudando o coração, a mente e as calças. A arqueologia sensorial [Changing the heart, the mind and the pants. Sensory Archeology]. Revista de Arqueologia do Museu de Arqueologia e Etnologia, São Paulo 20:3–16

Pellini (2016) Arqueologia e os sentidos. Entrando na toca do coelho [Archeology and the senses. Entering the rabbit hole]. Prismas

Peris J, Felicísimo A, Polo M (2013) Three-dimensional models of archaeological objects: from laser scanners to interactive PDF documents. In: Technical briefs in historical archaeology, vol 7, pp 13–18

Pett D, Bonacchi C (2012) Conclusions. In: Bonacchi, C. (orgs). Archaeology and digital communication. Towards strategies of public engagement. Archetype Publications. London, pp 126–130

Richardson L (2009) Measuring the success of digital public archaeology: towards a model for public participation in internet archaeology. Unpublished Master Thesis, University of London, London, 77p

Richardson L (2013) A digital public archaeology. Pap Inst Archaeol 23(1):1–12

Sabloff J (2008) Archaeology matters: action archaeology in the Modern World. Left Coast Press, Walnut Creek, California, U.S.A

Taylor J, Berggren A, Dell'unto N et al (2015) Revisiting reflexive archaeology at Çatalhöyük: integrating digital and 3D technologies at the trowel's edge. Antiquity 89:433–448

Tilley C (1994) The materiality of stone. Explorations in landscape phenomenology: 1. Berg. Oxford e New York

Tratado Antártico [Antarctic Treaty] (1959) Washington, December 1st., 1959

Tsiafaki D, Michailidou N (2015) Benefits and problems through the application of 3D technologies in archaeology: recording, visualization, representation and reconstruction. Sci Cult 1(3):37–45

Wolf E (1982) Europe and the people without History. University of California Press, Berkeley

Zarankin A, Senatore X (2007) Histórias de um passado em branco: arqueología historica Antártica [Histories of a White past: Antarctica historical archaeology]. Argumentum, Belo Horizonte, p 2007

Zarankin A, Hissa S, Salermo et al (2011) Paisagens em branco: arqueologia e antropologia antárticas – avanços e desafios [Blank landscpaes: Antarctica archaeology and anthropology – advances and challenges. Vestígios. Revista Latino-Americana de Arqueologia Histórica 5(2):9–52